Ben Stacy Jerrik (Ed.)

1948 Lutana Crash

Ben Stacy Jerrik (Ed.)

1948 Lutana Crash

Douglas DC-3, Australian National Airways, Australian National Airways, National aviation authority

Part Press

Imprint

Permission is granted to copy, distribute and/or modify this document under the terms of the GNU Free Documentation License, Version 1.2 or any later version published by the Free Software Foundation; with no Invariant Sections, with the Front-Cover Texts, and with the Back- Cover Texts. A copy of the license is included in the section entitled "GNU Free Documentation License".

All parts of this book are extracted from Wikipedia, the free encyclopedia (www.wikipedia.org).

You can get detailed informations about the authors of this collection of articles at the end of this book. The editors (Ed.) of this book are no authors. They have not modified or extended the original texts.

Pictures published in this book can be under different licences than the GNU Free Documentation License. You can get detailed informations about the authors and licences of pictures at the end of this book.

The content of this book was generated collaboratively by volunteers. Please be advised that nothing found here has necessarily been reviewed by people with the expertise required to provide you with complete, accurate or reliable information. Some information in this book maybe misleading or wrong. The Publisher does not guarantee the validity of the information found here. If you need specific advice (f.e. in fields of medical, legal, financial, or risk management questions) please contact a professional who is licensed or knowledgeable in that area.

Any brand names and product names mentioned in this book are subject to trademark, brand or patent protection and are trademarks or registered trademarks of their respective holders. The use of brand names, product names, common names, trade names, product descriptions etc. even without a particular marking in this works is in no way to be construed to mean that such names may be regarded as unrestricted in respect of trademark and brand protection legislation and could thus be used by anyone.

Cover image: www.ingimage.com
Concerning the licence of the cover image please contact ingimage.

Publisher:
Part Press is a trademark of
International Book Market Service Ltd., 17 Rue Meldrum, Beau Bassin, 1713-01 Mauritius
Email: info@bookmarketservice.com
Website: www.bookmarketservice.com

Published in 2012

Printed in: U.S.A., U.K., Germany. This book was not produced in Mauritius.

ISBN: 978-613-6-27651-9

Contents

Articles

References

1948_Lutana_crash

Occurrence summary	
Date	2 September 1948
Type	Controlled flight into terrain
Site	Near Nundle, New South Wales, Australia 31°30′40″S 150°55′59″E
Passengers	10
Crew	3
Fatalities	13
Aircraft type	Douglas DC-3
Operator	Australian National Airways
Tail number	VH-ANK
Flight origin	Sydney, New South Wales
Destination	Brisbane, Queensland

The **1948 *Lutana* crash** occurred on 2 September 1948 near Nundle, New South Wales, Australia, when the *Lutana*, a Douglas DC-3 operated by Australian National Airways en route to Sydney from Brisbane, crashed into high terrain due to navigation equipment errors, killing all 13 on board.

Flight

On 2 September 1948, the *Lutana* departed Brisbane's airport on a scheduled flight to Sydney. About 280 nautical miles (520 km) south of Brisbane it crashed into rising terrain in the North West Slopes of Australia's Great Dividing Range, due to an erroneously determined position based on errors in the navigational equipment the pilots relied upon for determining a safe course through the rising terrain.[1]

Investigation

The official inquiry was conducted by the Department of Civil Aviation. It consisted of a chairman, Judge William Simpson of the Supreme Court of the Australian Capital Territory, and two assessors, E. J. Bowen, Sci. D, Ph. D; and Captain L. M. Diprose, chief pilot of Associated Airlines, nominated by the Australian Pilots Association. The inquiry report, released 17 November 1948, found the pilot, Captain J. A. Drummond, to be a "pilot of more than ordinary ability," and led to a reorganization of the Department's system of air traffic control. The inquiry found that the probable cause of the crash was interference with the airplane's magnetic compass due to a nearby electrical storm and a defect in the navigational signals sent by the Government-maintained Kempsey low-frequency radio range station, an important navigational aid to flights in the area. The inquiry also identified errors and deficiencies in the aeronautical charts used to navigate the mountainous area.[1]

Australia's Air Minister, Mr. Drakeford, objected to the findings of the inquiry, stating that the lack of definitive evidence in the report rendered its findings "inconclusive," and that the assertion that the Kempsey range station malfunctioned was "difficult to believe."[1]

See also

- 1946 Australian National Airways DC-3 crash - Accident in Hobart, Tasmania
- 1949 MacRobertson Miller Aviation DC-3 crash - Accident in Perth, Western Australia

References

[1] "Court Says Directional Aids Misled Lutana's Pilot" (http://trove.nla.gov.au/ndp/del/article/22704590). *The Argus*: p. 7. 1948-11-25. . Retrieved 2011-09-19.

Douglas_DC-3

<table>
<tr><td colspan="2" align="center">DC-3</td></tr>
<tr><td colspan="2">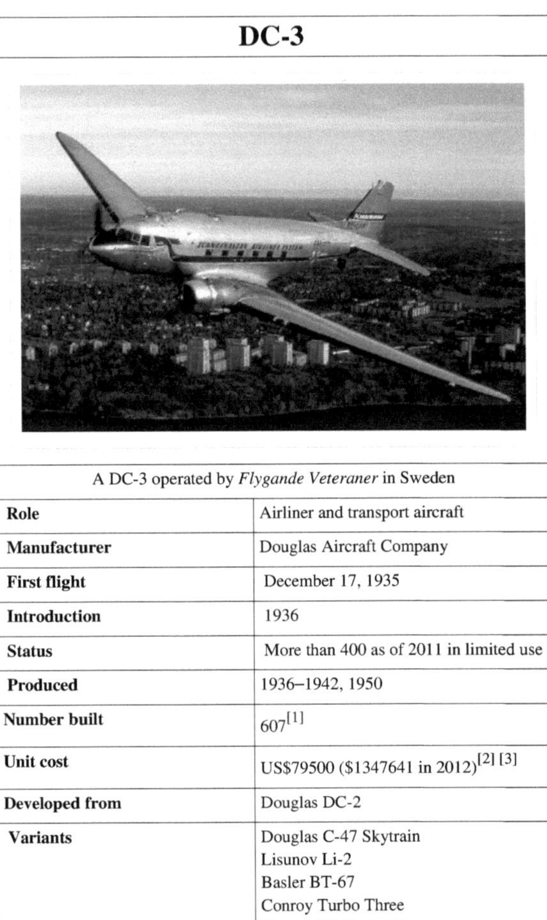
A DC-3 operated by Flygande Veteraner in Sweden</td></tr>
<tr><td>Role</td><td>Airliner and transport aircraft</td></tr>
<tr><td>Manufacturer</td><td>Douglas Aircraft Company</td></tr>
<tr><td>First flight</td><td>December 17, 1935</td></tr>
<tr><td>Introduction</td><td>1936</td></tr>
<tr><td>Status</td><td>More than 400 as of 2011 in limited use</td></tr>
<tr><td>Produced</td><td>1936–1942, 1950</td></tr>
<tr><td>Number built</td><td>607[1]</td></tr>
<tr><td>Unit cost</td><td>US$79500 ($1347641 in 2012)[2] [3]</td></tr>
<tr><td>Developed from</td><td>Douglas DC-2</td></tr>
<tr><td>Variants</td><td>Douglas C-47 Skytrain
Lisunov Li-2
Basler BT-67
Conroy Turbo Three
Conroy Tri-Turbo-Three</td></tr>
</table>

The **Douglas DC-3** is an American fixed-wing propeller-driven airliner, the speed and range of which revolutionized air transport in the 1930s and 1940s. Its lasting impact on the airline industry and World War II makes it one of the most significant transport aircraft ever made. The major military version was designated the C-47 Skytrain. Many DC-3/C-47s are still used in all parts of the world.

Design and development

The DC-3 was the culmination of a development effort that originated out of an inquiry from Transcontinental and Western Airlines (TWA) to Donald Douglas. TWA's rival in transcontinental air service, United Airlines, was inaugurating service with the Boeing 247 and Boeing refused to sell any 247s to other airlines until United's order for 60 aircraft had been filled.[4] TWA asked Douglas to design and build an aircraft that would enable TWA to compete with United. Douglas' resulting design, the 1933 DC-1, was promising, and led to the DC-2 in 1934. While the DC-2 was a success, there was still room for improvement.

The DC-3 was the result of a marathon telephone call from American Airlines CEO C. R. Smith to Donald Douglas, during which Smith persuaded a reluctant Douglas to design a sleeper aircraft based on the DC-2 to replace American's Curtiss Condor II biplanes. Douglas agreed to go ahead with development only after Smith informed him of American's intention to purchase twenty aircraft. The new aircraft was engineered by a team led by chief engineer Arthur E. Raymond over the next two years, and the prototype **DST** (for Douglas Sleeper Transport) first flew on December 17, 1935 (the 32nd anniversary of the Wright Brothers' flight at Kitty Hawk, North Carolina). A version with 21 passenger seats instead of the sleeping berths of the DST was also designed and given the designation **DC-3**. There was no prototype DC-3, the first DC-3 built followed seven DSTs off the production line and was delivered to American.[5]

The amenities of the DC-3 and DST popularized air travel in the United States. With only three refueling stops, eastbound transcontinental flights crossing the U.S. in approximately 15 hours became possible. Westbound trips took 17-1/2 hours due to prevailing headwinds — still a significant improvement over the competing Boeing 247. During an earlier era, such a trip would entail short hops in slower and shorter-range aircraft during the day, coupled with train travel overnight.[6]

A former military C-47B) of Air Atlantique taking off at RAF Hullavington.

A variety of radial engines were available for the DC-3 throughout the course of its development. Early-production civilian aircraft used Wright R-1820 Cyclone 9s, but later aircraft (and most military versions) used the Pratt & Whitney R-1830 Twin Wasp which offered better high-altitude and single engine performance, such as the three DC-3S *Super DC-3s* with Pratt & Whitney R-2000 Twin Wasps built in the late 1940s.

Production

Total production of all derivatives was 16,079.[7] More than 400 remained in commercial service in 1998. Production was as follows

607 civil variants of the DC-3.

10,048 military C-47 derivatives were built at Santa Monica, California, Long Beach, California, and Oklahoma City.

4,937 were built under license in Russia as the Lisunov Li-2 (NATO reporting name: **Cab**).

487 Mitsubishi Kinsei-engined aircraft were built by Showa and Nakajima in Japan, as the L2D2-L2D5 Type 0 transport. (Allied codename **Tabby**).

Production of civil DC-3s ceased in 1942, military versions were produced until the end of the war in 1945. In 1949, a larger, more powerful Super DC-3 was launched to positive reviews; however, the civilian market was flooded with second-hand C-47s, many of which were converted to passenger and cargo versions and only three were built and delivered the following year. The prototype Super DC-3 served the US Navy with the designation YC-129 alongside 100 C-47s that had been upgraded to the Super DC-3 specification.

Turboprop conversions

A BSAS C47–65ARTP powered by two Pratt & Whitney Canada PT6-65AR engines, formerly operated by the National Test Pilot School

From the early 1950s, some DC-3s were modified to use Rolls-Royce Dart engines, as in the Conroy Turbo Three. Other conversions featured Armstrong Siddeley Mamba and Pratt & Whitney PT6A turbines.

The Basler BT-67 is a conversion of the DC-3/C-47s. Basler refurbishes C-47/DC-3s at Oshkosh, Wisconsin, fitting them with Pratt & Whitney Canada PT6A-67R turboprop engines, lengthening the fuselage by 40 in (**unknown operator: u'strong'** cm) with a fuselage plug ahead of the wing and strengthening the airframe in selected areas. The airframe is rated as having "zero accumulated fatigue damage." This and extensive modifications to various systems and avionics result in a practically brand-new aircraft. The BT-67s have been supplied to civil and military customers in several countries.[8]

Braddick Specialised Air Services International PTY Ltd in South Africa is another company able to perform a Pratt & Whitney PT6 turboprop conversion of DC-3s. Over 50 DC-3/C-47s / 65ARTP / 67RTP / 67FTPs have been modified.[9]

Conroy Aircraft also made a three engine conversion with Pratt & Whitney Canada PT6 called the Conroy Tri-Turbo-Three.

Operational history

American Airlines inaugurated passenger service on June 26, 1936, with simultaneous flights from Newark, N.J. and Chicago, IL.[10] Early U.S. airlines like American, United, TWA and Eastern ordered over 400 DC-3s. These fleets paved the way for the modern American air travel industry, quickly replacing trains as the favored means of long-distance travel across the United States.

KLM Royal Dutch Airlines received its first DC-3 in 1936, it replaced the DC-2 on the service from Amsterdam via Batavia (now Jakarta) to Sydney, by far the longest scheduled route in the world at the time.

Douglas C-47B of Aigle Azur (France) in 1953, fitted with a ventral Turbomeca Palas booster jet for hot and high operations.

Piedmont Airlines operated DC-3/C-47s from 1948 to 1963. A DC-3 painted in the representative markings of Piedmont, operated by the Carolinas Aviation Museum, was retired from flight in March 2011. Both Delta Air Lines and Continental Airlines operate "commemorative" DC-3s wearing period markings.

During World War II, many civilian DC-3s were drafted for the war effort and just over 10,000 US military versions of the DC-3 were built, under the designations C-47, C-53, R4D, and Dakota. Peak production was reached in 1944, with 4,853 being delivered. The armed forces of many countries used the DC-3 and its military variants for the transport of troops, cargo, and wounded.

DC-3 on amphibious EDO floats. Sun-n-Fun 2003, Lakeland, Florida, United States

Licensed copies of the DC-3 were built in Japan as **Showa L2D** (487 aircraft) and in the USSR as the **Lisunov Li-2** (4,937 aircraft)[7]

Thousands of surplus C-47s, previously operated by several air forces, were converted for civilian use after the war and became the standard equipment of almost all the world's airlines, remaining in front line service for many years. The ready availability of cheap, easily maintained ex-military C-47s, both large and fast by the standards of the day, jump-started the worldwide post-war air transport industry. While aviation in pre-war Continental Europe had used the metric system, the overwhelming dominance of C-47s and other US war-surplus types cemented the use of nautical miles, knots and feet in post-war aviation throughout the world.

DC-3A that was used in the No.42 Squadron of the Royal New Zealand Air Force

Douglas had developed an improved version, the *Super DC-3*, with more engine power, greater cargo capacity, and a different wing but, with all the bargain-priced surplus aircraft available, this did not sell well in the civil aviation market. Only five were delivered, three of them to Capital Airlines. The U.S. Navy had 100 of their early R4Ds converted to Super DC-3 standard during the early 1950s as the R4D-8, later C-117D. The last U.S. Navy C-117 was retired July 12, 1976.[11] The last U.S. Marine Corps C-117, serial 50835, was retired from active service during June 1982. Several remained in service with small airlines in North and South America in 2006.[12]

A number of aircraft companies attempted to design a "DC-3 replacement" over the next three decades (including the very successful Fokker F27 Friendship), but no single type could match the versatility, rugged reliability, and economy of the DC-3. It remained a significant part of air transport systems well into the 1970s.

Douglas DC-3 today

A C-47A of Rovos Air in service in South Africa, 2006

There are still small operators with DC-3s in revenue service and as cargo aircraft. The common saying among aviation buffs and pilots is that "the only replacement for a DC-3 is another DC-3." The aircraft's legendary ruggedness is enshrined in the lighthearted description of the DC-3 as "a collection of parts flying in loose formation."[13] Its ability to take off and land on grass or dirt runways makes it popular in developing countries, where runways are not always paved.

Some of the uses of the DC-3 have included aerial spraying, freight transport, passenger service, military transport, missionary flying, and sport skydiving shuttling and sightseeing.

Perhaps unique among prewar and wartime aircraft, the DC-3 is in daily use. The very large number of civil and military operators of the DC-3/C-47s and related types, means that a listing of all the airlines, air forces and other current operators is impractical. As of 2012, DC-3 #10 is still used daily for flights in Colombia. [14]

The oldest surviving DC-3 is N133D, the sixth Douglas Sleeper Transport built in 1936. This aircraft was delivered to American Airlines on July 12, 1936 as NC16005. The aircraft was at Griffin-Spaulding County Airport, Griffin, Georgia as of November 2010, where it was being prepared for a ferry flight to Charlotte County Airport, Punta Gorda, Florida. The aircraft will be restored back to Douglas Sleeper Transport standards, and full airworthiness.[15]

The oldest DC-3 still flying is the original American Airlines *Flagship Detroit* (c/n 1920, #43 off the Santa Monica production line),[16] which can be seen at airshows around the United States and is owned and operated by the nonprofit Flagship Detroit Foundation.[17]

Indigo Aviation DC-3 before takeoff at Pemba Airport (Tanzania), August 2009

Variants

DST

> Douglas Sleeper Transport, the initial variant, 24 passengers during day and fitted out with 16 sleeper accommodation in the cabin for night.[18]

DC-3

> variant of DST with 21 passenger seats.

DC-3A

> Improved DC-3 with two 1,200 hp (895 kW) Pratt & Whitney R-1830-21 radial piston engines.

Fujairah Airlines DC-3 in the late 1960s

DC-3B

> Improved DC-3 with two Wright R-1820 Cyclone engines.

DC-3C

> Designation for ex-military C-47, C-53 and R4D aircraft sold on the civil market.[19]

DC-3S

> Super DC-3, improved DC-3 with a new wing, tail, and powered by two Pratt & Whitney R-2000 engines.

LXD1

> A single DC-3 supplied for evaluation by the Imperial Japanese Navy Air Service.

C-41A

> A single DC-3A (40-070) modified as a VIP transport, powered by two 1,200 hp (895 kW) Pratt & Whitney R-1830-21 radial piston engines, used to fly the Secretary of War.[20] (The Douglas C-41 was not a DC-3 derivative but a modification of a Douglas C-33.)

C-48

> One former United Air Lines DC-3A impressed.

C-48A

> Three impressed DC-3As with 18-seat interiors.

A captured Nakajima L2D in US markings, Mindanao, Philippines, May 1945

C-48B

Sixteen impressed former United Air Lines DST-As with 16-berth interior used as air ambulances.

C-48C

Sixteen impressed DC-3As with 21-seat interiors.

C-49

Various DC-3 and DST models, 138 impressed into service as C-49, C-49A, C-49B, C-49C, C-49D, C-49E, C-49F, C-49G, C-49H, C-49J, and C-49K.

C-50

Various DC-3 models, 14 impressed as C-50, C-50A, C-50B, C-50C and C-50D.

C-51

One aircraft ordered by Canadian Colonial Airlines impressed into service, had starboard-side door.

C-52

DC-3A aircraft with R-1830 engines, five impressed as C-52, C-52A, C-52B, C-52C and C-52D.

C-68

Two DC-3As impressed with 21-seat interiors.

C-84

One impressed DC-3B aircraft.

R4D-2

Two Eastern Air Lines DC-3s impressed into USN service as VIP transports, later designated **R4D-2F** and later **R4D-2Z**.

R4D-4

Ten impressed DC-3s for the US Navy

R4D-4R

Seven impressed DC-3s as staff transports for the US Navy.

R4D-4Q

Radar countermeasures version of R4D-4 for the US Navy.

Dakota II

RAF designation for impressed DC-3s

Conversions

DC-3/2000

DC-3/C-47 engine conversion done by Airtech Canada, first offered in 1987. Powered by two PZL ASz-62IT radials.[21]

Basler BT-67

DC-3/C-47 conversion with a stretched fuselage, strengthened structure, modern avionics, and powered by two Pratt & Whitney Canada PT-6A-67R.

Conroy Turbo Three

> One DC-3/C-47 converted by Conroy Aircraft with two Rolls-Royce Dart Mk. 510 turboprop engines.

Conroy Super-Turbo-Three

> Same as the Turbo Three but converted from a Super DC-3. One converted.

Conroy Tri-Turbo-Three

> One DC-3/C-47 converted by Conroy Aircraft with three Pratt & Whitney Canada PT-6A turboprops.

Conroy Tri-Turbo-Three at Farnborough Airshow in 1978.

USAC DC-3 Turbo Express

> A turboprop conversion by the United States Aircraft Corporation, fitting Pratt & Whitney Canada PT6A-45R turboprop engines with an extended forward fuselage to maintain center of gravity. First flight of the prototype conversion, (N300TX), was on July 29, 1982.[22]

BSAS C-47TP Turbo Dakota

> A South African C-47 conversion, by Braddick Specialised Air Services, with two Pratt & Whitney Canada PT6A-65R turboprop engines, revised systems, stretched fuselage and modern avionics for the South African Air Force.

Military and foreign derivatives

Douglas C-47

> Production military DC-3A variant.

Showa/Nakajima L2D

> 487 License built DC-3s and derivatives for the IJNAS.

Lisunov Li-2 / PS-84

> 4,937 DC-3 derivatives license-built in the USSR.

Accidents and incidents

Further information: List of accidents and incidents involving the DC-3

Specifications (DC-3A)

Data from McDonnell Douglas Aircraft since 1920[1]

General characteristics

- **Crew:** 2
- **Capacity:** 21–32 passengers
- **Length:** (19.7 m)
- **Wingspan:** (29.0 m)
- **Height:** (5.16 m)
- **Wing area: unknown operator: u'strong'** sq ft (91.7 m^2)
- **Empty weight: unknown operator: u','unknown operator: u','unknown operator: u',' (unknown operator: u'strong'unknown operator: u','kg)**
- **Gross weight: unknown operator: u'strong'unknown operator: u','lb (unknown operator: u',' kg)**
- **Powerplant:** 2 × Wright R-1820 Cyclone 9-cyl. air-cooled radial piston engine, **unknown operator: u','unknown operator: u','unknown operator: u',' (unknown operator: u'strong'unknown operator: u','kW)** each
- **Powerplant:** 2 × Pratt & Whitney R-1830-S1C3G Twin Wasp 14-cyl. air-cooled two row radial piston engine, **unknown operator: u','unknown operator: u','unknown operator: u',' (unknown operator: u'strong'unknown operator: u','kW)** each
- **Propellers:** 3-bladed Hamilton Standard 23E50 series

Performance

- **Maximum speed: unknown operator: u'strong'** kn; **unknown operator: u'strong'** km/h (230 mph) at 8,500 ft (2,590 m)
- **Cruise speed: unknown operator: u'strong'** kn; **unknown operator: u'strong'** km/h (207 mph)
- **Service ceiling: unknown operator: u','unknown operator: u','unknown operator: u',' (unknown operator: u'strong'unknown operator: u','m)**
- **Rate of climb: unknown operator: u','unknown operator: u','unknown operator: u',' (unknown operator: u'strong'unknown operator: u','m/s)**
- **Wing loading:** 25.5 lb/sq ft **(unknown operator: u'strong'** kg/m²)
- **Power/mass:** 0.0952 hp/lb (156.5 W/kg)

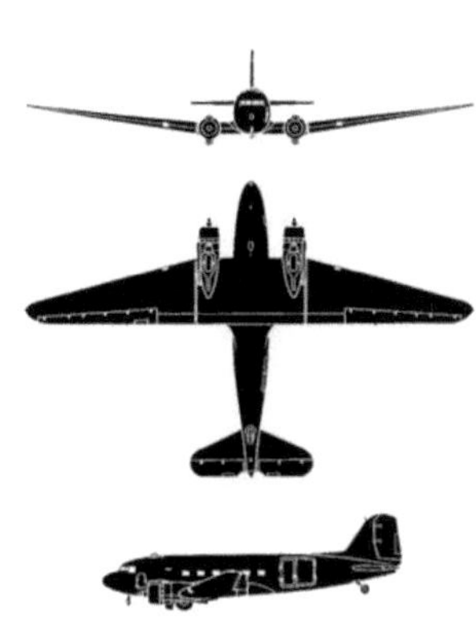

Douglas DC-3

Cockpit of DC-3 operated by FAA to verify operation of navaids (VORs and NDBs) along federal airways

See also

Related development

- Basler BT-67
- Douglas C-47 Skytrain
- Douglas AC-47 Spooky
- Lisunov Li-2
- Showa/Nakajima L2D

Aircraft of comparable role, configuration and era

- Boeing 247
- Curtiss-Wright C-46 Commando
- Junkers Ju 52
- Lockheed Model 18 Lodestar
- Saab 90 Scandia

Related lists

- List of aircraft of World War II
- List of civil aircraft

References

Notes

[1] Francillon 1979, pp. 217–251.
[2] Staff. Consumer Price Index (estimate) 1800–2012 (http://www.minneapolisfed.org/community_education/teacher/calc/hist1800.cfm). Federal Reserve Bank of Minneapolis. Retrieved February 22, 2012.
[3] Rumerman, Judy. "The Douglas DC-3" (http://www.centennialofflight.gov/essay/Aerospace/DC-3/Aero29.htm). U.S. Centennial of Flight Commission. . Retrieved March 12, 2012.
[4] O'Leary 1992, p. 7.
[5] Pearcy 1987, p. 17.
[6] O'Leary 2006, p. 54.
[7] Gradidge 2006, p. 20.
[8] "Basler BT-67." (http://www.baslerturbo.com/specifications.html) *Basler Turbo Conversions, LLC* via *baslerturbo.com*, 2008. Retrieved: March 7, 2009.
[9] "BSAS International." (http://www.bsasinternational.com/) *www.bsasinternational.com*. Retrieved: October 11, 2011.
[10] Holden, Henry. "DC-3 History." (http://www.dc3history.org/douglasdc3.html) *dc3history.org*. Retrieved: October 7, 2010.
[11] "The Seventies 1970–1980: C-117, p. 316." (http://www.history.navy.mil/avh-1910/PART10.PDF) *history.navy.mil*. Retrieved: August 10, 2010.
[12] Gradidge 2006, pp. 634–637.
[13] Williams, Michael. "How health and safety rules have grounded the Dakota, the war workhorse." (http://www.dailymail.co.uk/news/article-518349/How-health-safety-rules-grounded-Dakota-war-workhorse.html) *Daily Mail*, February 25, 2008. Retrieved: March 7, 2009.
[14] "Colombia's Workhorse, the DC-3 airplane." (http://www.washingtonpost.com/world/the_americas/colombias-workhorse-the-dc-3-airplane/2012/03/09/gIQA8yN36R_gallery.html#photo=1) *Washington Post*. Retrieved: March 15, 2012.
[15] McSwiggen, Bob. "World's Oldest DC-3." (http://www.douglasdc3.com/olddc3/olddc3.htm) *douglasdc3.com*, 2011. Retrieved: August 9, 2011.
[16] Pearcy 1985
[17] "DC-3." (http://www.flagshipdetroit.org/) *Flagship Detroit Foundation*. Retrieved: October 7, 2010.
[18] "Sleeping Car of the Air Has Sixteen Sleeping Berths." (http://books.google.com/books?id=QdsDAAAAMBAJ&pg=PA23&dq=Popular+Science+1935+plane+"Popular+Mechanics"&hl=en&ei=QIs_TpjpHOPJsQKo4uC_Bw&sa=X&oi=book_result&ct=result&resnum=2&ved=0CCwQ6AEwATgU#v=onepage&q&f=true) *Popular Mechanics*, January 1936.
[19] "Aircraft Specifications NO. A-669." (http://www.douglasdc3.com/faa/a-669.pdf) *FAA*. Retrieved: October 20, 2011.
[20] "Douglas C-41A." (http://www.aero-web.org/specs/douglas/c-41.htm) *aero-web.org*. Retrieved: August 10, 2010.
[21] "AirTech Company Profile." (http://www.ic.gc.ca/app/ccc/srch/nvgt.do?lang=eng&prtl=1&sbPrtl=&estblmntNo=900434230000&profile=cmpltPrfl&profileId=801&app=sold) *ic.gc.ca*. Retrieved: November 22, 2009.
[22] Taylor 1983

Bibliography

- Francillon, René. *McDonnell Douglas Aircraft Since 1920: Volume I*. London: Putnam, 1979. ISBN 0-87021-428-4.
- Gradidge, Jennifer M. *The Douglas DC-1/DC-2/DC-3: The First Seventy Years Volumes One and Two*. Tonbridge, Kent, UK: Air-Britain (Historians) Ltd., 2006. ISBN 0-85130-332-3.
- O'Leary, Michael. *DC-3 and C-47 Gooney Birds*. St. Paul, Minnesota: Motorbooks International, 1992. ISBN 0-87938-543-X.
- O'Leary, Michael. "When Fords Ruled the Sky (Part Two)." *Air Classics*, Volume 42, No. 5, May 2006.
- Pearcy, Arthur. *Douglas DC-3 Survivors, Volume 1*. Bourne End, Bucks, UK: Aston Publications, 1987. ISBN 0-946627-13-4.
- Pearcy, Arthur. *Douglas Propliners: DC-1–DC-7*. Shrewsbury, UK: Airlife Publishing, 1995. ISBN 1-85310-261-X.
- Taylor, John W. R. *Jane's All the World's Aircraft, 1982–83*. London: Jane's Publishing Company, 1983. ISBN 0-7106-0748-2.
- Yenne, Bill. *McDonnell Douglas: A Tale of Two Giants*. Greenwich, Connecticut: Bison Books, 1985. ISBN 0-517-44287-6.

External links

- DC-3/Dakota Historical Society (http://www.dc3history.org)
- Buffalo Airways DC3 Specifications, as seen on [[Ice Pilots NWT (http://www.buffaloairways.com/fleet/douglas-dc-3/)] / Discovery Channel]
- Dutch Dakota Association operates two flying Douglas DC-3s, part of Aviodrome Aviation Museum Lelystad-Holland (http://www.dutchdakota.nl/)
- The DC-3 Hangar – Douglas DC-3 specific site (http://www.douglasdc3.com/)
- The DC-3 Aviation Museum. A registered non-profit organization working to preserve the history of the DC-3 (http://www.dc3museum.org/)
- Centennial of flight Commission on the DC-3 (http://www.centennialofflight.gov/essay/Aerospace/DC-3/Aero29.htm)
- Aviation history on the DC-3 (http://www.aviation-history.com/douglas/dc3.html)
- The Flagship Detroit Foundation (http://www.flagshipdetroit.org)
- Aviation history on the DC-3 (http://www.aviation-history.com/douglas/dc3.html)
- Flying Piedmont Airlines DC-3 (http://www.carolinasaviation.org/collections/aircraft/dc3n44v.html), formerly owned and operated by Piedmont Airlines, and US Air
- Basler BT-67 Re-manufactured and turbinized version of the DC-3 (http://www.baslerturbo.com/specifications.html)
- Warbird Alley: DC-3/C-47 page (http://www.warbirdalley.com/c47.htm)
- Early 1970s DC-3 turboprop conversions by Conroy Aircraft (http://www.air-and-space.com/conroy.htm)
- Gold Timer Foundation: Restorers of the only remaining Li-2 still in flying condition. HA-LIX flies sight-seeing flights. (Hungarian) (http://www.goldtimer.hu/)
- Dakota Norway (http://www.dakotanorway.no/)
- DC-Association Finland (http://www.dc-ry.fi/)
- Link for a number of South African DC-3 flights (http://www.historicflight.co.za/)
- Interview with DC-3 pilot and instructor and discussion of what it takes to become type-rated in the DC-3 (http://media.libsyn.com/media/airspeed/AirspeedGryder.mp3)
- Audio that follows a training program for a US second-in-command type rating in the DC-3. Includes cockpit audio (http://media.libsyn.com/media/airspeed/AirspeedDC-3Summary.mp3)
- CNAPG Douglas Dakota production list (http://www.cnapg.net/dakota.htm)

- The Dakota Association of South Africa (http://www.dc-3.co.za/)
- The South African Airways Museum Society Owners of Douglas DC-3 ZS-BXF (http://www.saamuseum.co.za/)
- The site about the last Douglas DC-3 flying in France (F-AZTE) – Francedc3.com (http://www.francedc3.com/)
- Ohio University's 1943 Douglas DC-3 page (http://www.ohio.edu/avionics/aircraft/1943douglasdc-3.cfm/)

Australian_National_Airways

Australian National Airways

[[Image:ANA logo AA.png			alt=]]
IATA	**ICAO**	**Callsign**	
Founded	1936		
Commenced operations	1936		
Ceased operations	1957		
Destinations			
Company slogan	*"Wing your way with ANA"*		
Headquarters			

Australian National Airways (ANA) was Australia's predominant carrier from the mid-1930s to the early 1950s.

The Holyman Airways Period

On 19 March 1932 **Flinders Island Airways** began a regular aerial service using the Desoutter Mk.II VH-UEE *Miss Launceston* between Launceston, Tasmania and Flinders Island in Bass Strait, which competed with shipping services offered by William Holyman and Sons Ltd. Due to monopoly arrangements with other Australian shipowners, Holymans (as it was known) was only allowed to carry passengers on internal Tasmanian routes, and resented the intrusion. Brothers Captain Victor Holyman and Ivan Holyman purchased a de Havilland D.H.83 Fox Moth VH-UQM *Miss Currie* which entered service on the same route on 1 October 1932, and soon amalgamated with Flinders Island Airways to form **Tasmanian Aerial Services Pty. Ltd**. They later purchased a de Havilland D.H.84 Dragon VH-URD *Miss Launceston* that began a regular service between Melbourne, Flinders Island and Launceston in September 1933.

Following the Australian Government's announcement of the Empire Air Mail Scheme late in 1933, Holymans entered into a partnership with the two main shipping companies servicing Tasmania, Huddart Parker and the Union Steamship Company of New Zealand, to form an equal-share partnership in a new **Holyman's Airways Pty. Ltd.** headed by Ivan Holyman. The new company, with a capital of £90,000, was registered in July 1934, and ordered two de Havilland D.H.86 Express airliners. The first of these, VH-URN *Miss Hobart*, began operating across Bass Strait on 28 September 1934, but went missing just three weeks later, on 18 October, and was believed to have crashed off Wilsons Promontory. Captain Victor Holyman's was one of the twelve lives lost.

Douglas DC-2 VH-USY *Bungana* in 1936.

Undaunted, Holyman's Airways purchased a second-hand D.H.84 (VH-URG *Golden West*) and ordered two more D.H.86s, and soon began to expand operations throughout south-eastern Australia. A route from Melbourne to Sydney via Canberra was established in 1935 using D.H.86 VH-UUB *Loila*. On the day of a first proving flight between the capitals, 2 October, another D.H.86, VH-URT *Loina*, crashed into Bass Strait off Flinders Island killing all five on board. The Melbourne-Sydney flights, the first regular daily airmail service between the two centres, got underway on 7 October.

After a non-fatal accident in Bass Strait to the D.H.86 VH-USW *Lepena* on 13 December 1935, Ivan Holyman used his influence with the Australian Government to have an official ban on the importation of US-built commercial aircraft to be lifted, and Holyman's Airways ordered an example of the recently introduced Douglas DC-2. It entered service as VH-USY *Bungana* on 18 May 1936.

Restructure as Australian National Airways

Early in 1936 Ivan Holyman approached the Adelaide Steamship Company, owners of Adelaide Airways, with a view to an amalgamation aiming to become Australia's most powerful airline. Adelaide Airways had recently taken over West Australian Airways and the new combine would thus effectively control airline traffic between Perth, Adelaide, Melbourne and Sydney. With funding from the Orient Steam Navigation Company a new **Australian National Airways** was registered on 13 May 1936, and began services under its new name on 1 July 1936. It acquired a second DC-2 VH-UXJ *Loongana* that began a twice-weekly service between Melbourne and Perth on 21 December 1936.

Meanwhile efforts to expand operation northwards to Queensland were being thwarted by Airlines of Australia (AoA), its main competitor. Established in 1931 as **New England Airways** by G.A. Robinson and Keith Virtue of Lismore, it operated services in northern New South Wales and between Sydney and Brisbane, Queensland, expanding further into Queensland by taking over a number of struggling regional airlines during the mid 1930s. It was restructured as AoA in 1934 with funding by an investment group the British Pacific Trust. In 1936 it introduced Stinson Model A airliners in a regular service between Sydney and Brisbane, and later acquired Douglas DC-2s and Douglas DC-3s. After several months of fruitless negotiations with its financiers, ANA managed to gain a controlling interest in AoA in April 1937, although the two airlines retained separate public identities until 1942. Between them the two airlines operated four DC-2s and four DC-3s by the time of the outbreak of World War II, as well as several other aircraft including two Model As, two D.H.84s, two D.H.86s and nine de Havilland D.H.89 Rapides.

When Australia entered World War II in 1939 the Government of Australia requisitioned ANA's four DC-3s, leaving it to battle on with its assortment of lesser aircraft. However, ANA was soon operating a network of services around Australia on behalf of the war effort. It operated a large number of Douglas DC-2s, DC-3s and even at least one rare Douglas DC-5, mostly on the behalf of the American forces in Australia.

Notable Accidents

During the 1940s ANA was plagued by a series of accidents and disasters that resulted in considerable adverse publicity. The most serious of these were:

- 25 October 1938, DC-2 VH-UYC *Kyeema* overflew Essendon Airport and crashed into Mount Dandenong. All four crew and fourteen passengers were killed.[1]

- 8 February 1940, DC-2 VH-USY crash-landed near Dimboola after engine fire. No loss of life and aircraft repaired.

- 29 May 1942, DH-89 VH-UXZ *Marika* crashed off Flinders Island Bass Strait apparently after running out of fuel, all four on board drowned.

Australian National Airways Douglas DC-4 VH-ANA *Amana*

- 3 December 1943, DC-2 VH-ADQ crash landed near Bendigo after the pilot lost his way on a flight from Sydney to Melbourne — the first officer was killed but the aircraft was repaired.

- 31 January 1945, Stinson Model A VH-UYY *Tokana* broke up in mid-air due to metal fatigue of a wing joint and crashed near Heathcote, Victoria, killing all ten on board.

- 10 March 1946. In the Seven-Mile Beach crash, DC-3 VH-AET plunged into the sea shortly after taking off from Cambridge Aerodrome, killing all 25 on board. An Air Court of Inquiry examined a number of theories that might explain the accident but found there was insufficient evidence to determine any one as the cause. [2] [3]

- 2 September 1948, DC-3 VH-ANK *Lutana* crashed into high terrain near Nundle, NSW, due to navigation equipment errors, killing all 13 on board.

- 4 October 1948, DC-3 VH-ABR *Kanana* crash-landed near NW of Yass, New South Wales after an engine fire. No serious injuries and aircraft rebuilt (still flying in 2008).

- 8 November 1948, DC-3 VH-UZK *Kurana* on Mount Macedon Victoria after an unauthorised change of course by the captain, who was killed along with his first officer. Miraculously the hostess and nineteen passengers survived.

- 29 December 1948, VH-UZJ *Kyilla* crash-landed near Mangalore, Victoria after the pilot misjudged his altitude [this was proven to be incorrect when appealed by the pilot in court, the altimeter had failed and due to being night and poor weather conditions the pilot was relying on instruments to fly - the two altimeters in the aircraft read 500 feet and 450 feet when the pilot checked at the time of the crash] while landing and touched the ground at high speed. No serious injuries, but aircraft damaged beyond repair. [Please refer to pilot's own memoirs which can be found through http://www.trove.nla.gov.au and many newspaper articles that verify that Captain Harold Shelton won his appeal against A.N.A.]

Many of these accidents were put down to human error (generally on the part of the pilots), and a tightening of operational policies seemed to have arrested the problem. A final disaster was:

- 26 June 1950, DC-4 VH-ANA *Amana* crashed near York, Western Australia killing all 28 on board (a passenger survived the immediate crash but later died of his injuries). The cause was probably the result of fuel starvation as the result of a fuel tank water drainage port not being closed on the tarmac. The aircraft lost power on most engines and the crew had only just managed to stabilise fuel supply by isolating the "offending" cross-flow lines and regain power when they ran out of altitude.[4] The crash of the company's flagship was a blow to ANA's prestige and contrasted dramatically with its rival TAA's record which, at that stage, was unblemished.

Post-War Activities

Ben Chifley's Labor government was determined that post-war aviation would be a state monopoly. Its legislation was stymied, however, by the Airline Operators Secretariat, which argued that the Constitution guaranteed freedom of commerce between states. In its *Airlines Case* decision, the High Court agreed.

Although ANA had won, it now faced severe competition in the form of the state-owned airline Trans Australia Airlines. ANA had hitherto enjoyed a near-monopoly on domestic air transport. From the start, TAA was a better run airline. It particularly made better choices of aircraft than ANA. Ivan Holyman stuck to his relationship with Douglas, buying Douglas DC-4s and Douglas DC-6Bs, while TAA opted for Convair 240s and Vickers Viscounts.

By the mid-1950s TAA had driven ANA close to collapse. Holyman had wanted to expand overseas but the government's ownership of Qantas prevented this. He bought shareholdings in Cathay Pacific Airways and Air Ceylon, but ANA aircraft were never seen on international routes. In 1952 the conservative Menzies government declined to close TAA down, instead it provided ANA with finance to upgrade its fleet to compete with TAA. At this point Holyman opted for DC-6Bs while TAA went for the more attractive Viscount.

When Ivan Holyman died in 1957 the shareholders offered to sell out to the government and let ANA merge with TAA. The government declined.

ANA Douglas DC-4 aircraft at Perth Airport in 1955.

Takeover by Ansett Transport Industries (ATI)

After initially dismissing his offer, the ANA board began talking with Reginald Ansett, head of the much smaller Ansett Transport Industries Ltd; with its main interstate operation Ansett Airways. Finally, ANA was sold to Ansett, on 3 October 1957, for 3.3 million pounds. The two airlines were merged to form Ansett-ANA on 21 October 1957 and the name was retained until 1 November 1968 when it was renamed Ansett Airlines of Australia.

References

[1] Job, Macarthur (December 2008). "From Disaster to New Dawn". *Aeroplane*: p28–32. ISSN 0143-7240.

[2] Aussieairliners (http://www.aussieairliners.org/dc-3/vh-aet/vhaet.html) Retrieved 2011-09-14

[3] Aviation Safety Network (http://aviation-safety.net/database/record.php?id=19460310-0) Retrieved 2011-09-14

[4] "Skymaster Accident Summary" (http://aviation-safety.net/database/record.php?id=19500626-0&lang=en). Flight Safety Foundation. 21 February 2005. . Retrieved 2008-09-24.

• Yule, Dr. Peter, *The Forgotten Giant of Australian Aviation : Australian National Airways*, Hyland House, Melbourne 2001.

• Macarthur Job, *Air Crash - The Story of How Australia's Airways Were Made Safe*, Volumes I & II, Aerospace Publications, Weston Creek (Canberra), 1991, 1992.

National_aviation_authority

The **National Aviation Authority** (**NAA**) is the government statutory authority in each country that oversees the approval and regulation of civil aviation.

Role

Due to the inherent dangers in the use of flight vehicles, NAA's typically regulate the following critical aspects of aircraft airworthiness and their operation:

- Design of aircraft, engines, airborne equipment and ground-based equipment affecting flight safety
- Conditions of manufacture and test of aircraft and equipment
- Maintenance of aircraft and equipment
- Operation of aircraft and equipment
- Licensing of pilots and maintenance engineers
- Licensing of airports and navigational aids
- Standards for air traffic control

Depending on the legal system of the parent country, the NAA will derive its power from an act of Parliament (such as the Civil or Federal Aviation Act), and is then empowered to make regulations within the bounds of the act. This allows technical aspects of airworthiness to be dealt with by subject matter experts and not politicians.[1] [2]

The NAA may also be involved in the investigation of aircraft accidents, although in many cases this is left to a separate body (such as the Australian Transport Safety Bureau (ATSB) in Australia or the National Transportation Safety Board (NTSB) in the USA), to allow independent review of regulatory oversight.[3].

The NAA will regulate the control of air traffic but a separate agency will generally carry out Air Traffic Control functions.

History

The independent development of NAAs has resulted in differing regulations in country to country. This has required aircraft manufacturers in the past to develop differing models for specific NAA requirements (such as the BAe Jetstream 31), and difficulty for airlines to travel into foreign jurisdictions. In an effort to resolve these issues, the Convention on International Civil Aviation (Chicago Convention) was signed in 1944. This then led to the establishment by the United Nations established the International Civil Aviation Organization (ICAO) in 1947 which now oversees member states and works to implement regulatory changes to ensure best practice regulations are adopted.[4]

Major national aviation authorities

- Federal Aviation Administration (FAA, USA)
- Civil Aviation Safety Authority (CASA, Australia)
- Transport Canada (TC, Canada)
- Agência Nacional de Aviação Civil (ANAC, Brazil)
- Direction Générale de l'Aviation Civile (DGAC, France)
- Luftfahrt-Bundesamt (LBA, Germany)
- Civil Aviation Authority (United Kingdom) (CAA, UK)
- Civil Aviation Authority of New Zealand (CAA, NZ)
- European Aviation Safety Agency (EASA, is not actually an NAA but plays part of the role within its member states of the EU)

- Directorate General of Civil Aviation (DGCA, India)
- Civil Aviation Authority of Pakistan (CAA, Pakistan)
- Civil Aviation Administration of China (CAAC, People's Republic of China)
- Italian Civil Aviation Authority (ENAC, Italy)
- Civil Aviation Department (CAD, Hong Kong)
- Civil Aviation Authority of Singapore (CAAS, Singapore)
- Civil Aviation Authority of Turkey- Sivil Havacilik Genel Mudurlugu (SHGM, Turkey)

See also

- Air route authority between the United States and the People's Republic of China
- List of civil aviation authorities

References

[1] Australian Civil Aviation Act, 1988 (http://austlii.law.uts.edu.au/au/legis/cth/consol_act/caa1988154/)
[2] Federal Aviation Act of 1958, Embry-Riddle Aeronautical University Library (http://libraryonline.erau.edu/online-full-text/books-online/ Aviationlawpt1.pdf)
[3] About the Australian Transport Safety Bureau (ATSB) (http://www.atsb.gov.au/about_atsb/about.aspx)
[4] ICAO ICAO'S Aims (http://www.icao.int/icao/en/aimstext.htm)

Supreme_Court_of_the_Australian_Capital_Territory

<table>
<tr><td colspan="2" align="center">Supreme Court of the Australian Capital Territory</td></tr>
<tr><td>Established</td><td>1 January 1934</td></tr>
<tr><td>Jurisdiction</td><td>Australian Capital Territory, Australia</td></tr>
<tr><td>Location</td><td>Canberra</td></tr>
<tr><td>Coordinates</td><td>35°16′52″S 149°07′38″E</td></tr>
<tr><td>Decisions are appealed to</td><td>High Court of Australia</td></tr>
<tr><td>Judge term length</td><td>mandatory retirement by age of 70</td></tr>
<tr><td>Number of positions</td><td>4</td></tr>
<tr><td>Website</td><td>Supreme Court of the Australian Capital Territory [1]</td></tr>
<tr><td colspan="2" align="center">Chief Justice of the Australian Capital Territory</td></tr>
<tr><td>Currently</td><td>Terence Higgins</td></tr>
<tr><td>Since</td><td>31 January 2003</td></tr>
</table>

The **Supreme Court of the Australian Capital Territory** is the superior court for the ACT. It has unlimited jurisdiction within the territory in civil matters (although it usually only hears matters involving more than A$50,000), and hears the most serious criminal matters. In the Australian court hierarchy it is one of 8 state and territory Supreme Courts having unlimited jurisdiction in their respective parts of Australia.

The court was established on 1 January 1934 by the *Seat of Government Supreme Court Act 1933 (Cth)*. The first judge of the court was Lionel Lukin. The first Chief Judge was Russell Walter Fox, appointed when the office was created in 1977. It was substituted with the office of Chief Justice in 1982. The first Chief Justice was Richard Blackburn.

Pursuant to section 37E of the Supreme Court Act 1933 [2], the court is known as the Court of Appeal when it exercises its appellate jurisdiction under Part 2A of the Act.

Current judges

- Chief Justice Terence Higgins (31 January 2003)
- Justice Richard Refshauge (1 February 2008)
- Justice Hilary Penfold (1 February 2008)
- Justice John Burns (1 August 2011)

See also

- List of Judges of the Supreme Court of the Australian Capital Territory

References

- "The Supreme Court of the ACT" [1]. *ACT Law Courts*. Retrieved 1 June 2005.

Law Courts of the Australian Capital Territory
building

References

[1] http://www.courts.act.gov.au/supreme/default.asp?textonly=no
[2] http://www.legislation.act.gov.au/a/1933-34/current/pdf/1933-34.pdf

Compass

A **compass** is a navigational instrument that measures directions in a frame of reference that is stationary relative to the surface of the earth. The frame of reference defines the four *cardinal directions* (or *points*) – north, south, east, and west. Intermediate directions are also defined. Usually, a diagram called a compass rose, which shows the directions (with their names usually abbreviated to initials), is marked on the compass. When the compass is in use, the rose is aligned with the real directions in the frame of reference, so, for example, the "N" mark on the rose really points to the north. Frequently, in addition to the rose or sometimes instead of it, angle markings in degrees are shown on the compass. North corresponds to zero degrees, and the angles increase clockwise, so east is 90 degrees, south is 180, and west is 270. These numbers allow the compass to show azimuths or bearings, which are commonly stated in this notation.

A simple dry magnetic pocket compass

There are two widely used and radically different types of compass. The magnetic compass contains a magnet that interacts with the earth's magnetic field and aligns itself to point to the magnetic poles. The gyro compass (sometimes spelled with a hyphen, or as one word) contains a rapidly spinning wheel whose rotation interacts dynamically with the rotation of the earth so as to make the wheel precess, losing energy to friction until its axis of rotation is parallel with the earth's.

The magnetic compass was invented during the Chinese Han Dynasty between the 2nd century BC and 1st century AD,[1] and was used for navigation by the 11th century.[2] The compass was introduced to medieval Europe 150 years later,[2] where the dry compass was invented around 1300.[3] This was supplanted in the early 20th century by the liquid-filled magnetic compass.[4]

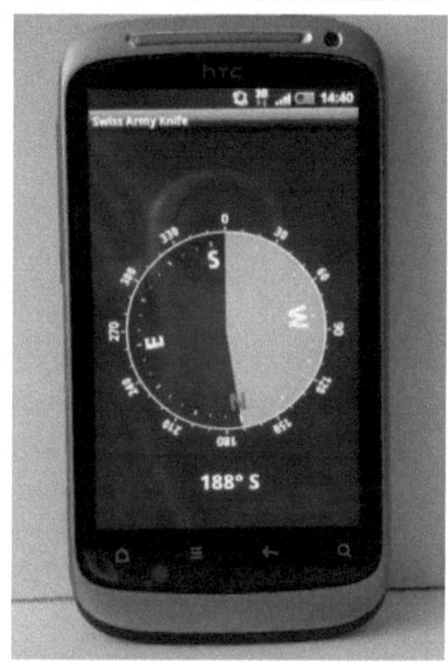
A smartphone that can be used as a compass because of the magnetometer inside

Types of compasses

There are two widely used and radically different types of compass. The magnetic compass contains a magnet that interacts with the earth's magnetic field and aligns itself to point to the magnetic poles.[5] Simple compasses of this type show directions in a frame of reference in which the directions of the magnetic poles are due north and south. These directions are called *magnetic north* and *magnetic south*. The gyro compass (sometimes spelled with a hyphen, or as one word) contains a rapidly spinning wheel whose rotation interacts dynamically with the rotation of the earth so as to make the wheel precess, losing energy to friction until its axis of rotation is parallel with the earth's. The wheel's axis therefore points to the earth's rotational poles, and a frame of reference is used in which the directions of the rotational poles are due north and south. These directions are called *true north* and *true south*, respectively.

There are other devices which are not conventionally called compasses but which do allow the true cardinal directions to be determined. They are said to work "like a compass", or "as a compass", without actually being a compass. For example, a Global Positioning System (GPS) satellite receiver determines its own position on the ground, as true latitude and true longitude. If the receiver is being moved, even at walking pace, it can follow the change of its position, and hence determine the compass bearing of its direction of movement, and thence the directions of the cardinal points relative to its direction of movement. Some GPS receivers have two antennas, fixed some distance apart to the structure of a vehicle, usually an aircraft. The exact latitudes and longitudes of the antennas can be determined simultaneously, which allows the directions of the cardinal points to be calculated relative to the heading of the aircraft (the direction in which its nose is pointing), rather than to its direction of movement, which will be different if there is a crosswind. A much older example was the Chinese south-pointing chariot, which worked like a compass by directional dead reckoning. It was initialized by hand, possibly using astronomical observations e.g. of the Pole Star, and thenceforth counteracted every turn that was made to keep its pointer aiming in the desired direction, usually to the south.

The earth's magnetic poles do not coincide with the rotational poles, and the positions of the magnetic poles change over time on a time-scale that is not extremely long by human standards. Significant movements happen in a few years. (Over millions of years, the directions of the true poles also shift, because of continental drift.) For an observer at any point on the earth's surface, there is an angle, called the *magnetic declination* (or *magnetic variation*), between the directions of magnetic north and true north. The magnetic declination is different at different points on the earth, and changes with time. Close to the equator, the magnetic declination is no more than a few degrees, but in arctic and Antarctic latitudes it can be much greater. Some magnetic compasses include means to compensate for the magnetic declination, so that the compass shows true directions, relative to the earth's rotational poles. The user of such a compass has to know the local value of the magnetic declination, and adjust the compass accordingly.

Magnetic compass

The magnetic compass consists of a magnetized pointer (usually marked on the North end) free to align itself with Earth's magnetic field. A compass is any magnetically sensitive device capable of indicating the direction of the magnetic north of a planet's magnetosphere. The face of the compass generally highlights the cardinal points of north, south, east and west. Often, compasses are built as a stand alone sealed instrument with a magnetized bar or needle turning freely upon a pivot, or moving in a fluid, thus able to point in a northerly and southerly direction.

The compass greatly improved the safety and efficiency of travel, especially ocean travel. A compass can be used to calculate heading, used with a sextant to calculate latitude, and with a marine chronometer to calculate longitude. It thus provides a much improved navigational capability that has only been recently supplanted by modern devices such as the Global Positioning System (GPS).

The compass was invented during the Chinese Han Dynasty between the 2nd century BC and 1st century AD.[1] The dry compass was invented in medieval Europe around 1300.[3] This was supplanted in the early 20th century by the liquid-filled magnetic compass.[4]

Other, more accurate devices have been invented for determining north that do not depend on the Earth's magnetic field for operation (known in such cases as true north, as opposed to magnetic north). A gyrocompass or astrocompass can be used to find true north, while being unaffected by stray magnetic fields, nearby electrical power circuits or nearby masses of ferrous metals. A recent development is the electronic compass, either a magnetometer or a fibre optic gyrocompass, which detects the magnetic directions without potentially fallible moving parts. A magnetometer frequently appears as an optional subsystem built into hand-held GPS receivers. However, magnetic compasses remain popular, especially in remote areas, as they are relatively inexpensive, durable, and require no power supply.[6]

How a magnetic compass works

A compass functions as a pointer to "magnetic north" because the magnetized needle at its heart aligns itself with the lines of the Earth's magnetic field. The magnetic field exerts a torque on the needle, pulling one end or *pole* of the needle toward the Earth's North magnetic pole, and the other toward the South magnetic pole. The needle is mounted on a low-friction pivot point, in better compasses a jewel bearing, so it can turn easily. When the compass is held level, the needle turns until, after a few seconds to allow oscillations to die out, one end points toward the North magnetic pole.

A magnet or compass needle's "north" pole is defined as the one which is attracted to the North magnetic pole of the Earth, in northern Canada. Since opposite poles attract ("north" to "south") the North magnetic pole of the Earth is actually the *south* pole of the Earth's magnetic field.[7] [8] [9] The compass needle's north pole is always marked in some way: with a distinctive color, luminous paint, or an arrowhead.

Instead of a needle, professional compasses usually have bar magnets glued to the underside of a disk pivoted in the center so it can turn, called a "compass card", with the cardinal points and degrees marked on it. Better compasses are *"liquid-filled"*; the chamber containing the needle or disk is filled with a liquid whose purpose is to damp the oscillations of the needle so it will settle down to point to North more quickly, and also to protect the needle or disk from shock.

In navigation, directions on maps are expressed with reference to *geographical* or *true north*, the direction toward the Geographical North Pole, the rotation axis of the Earth. Since the Earth's magnetic poles are near, but are not at the same locations as its geographic poles, a compass does not point to true north. The direction a compass points is called *magnetic north*, the direction of the North magnetic pole, located in northeastern Canada. Depending on where the compass is located on the surface of the Earth the angle between true north and magnetic north, called *magnetic declination* can vary widely, increasing the farther one is from the prime meridian of the Earth's magnetic field. The local magnetic declination is given on most maps, to allow the map to be oriented with a compass parallel to true north.

In geographic regions near the magnetic poles, in northeastern Canada and Antarctica, variations in the Earth's magnetic field cause magnetic compasses to have such large errors that they are useless, so other instruments must be used for navigation.

History

The first compasses were made of lodestone, a naturally magnetized ore of iron. Ancient Chinese people found that if a lodestone was suspended so it could turn freely, it would always point in the same direction (toward the magnetic poles). Later compasses were made of iron needles, magnetized by stroking them with a lodestone.

Navigation prior to the compass

Prior to the introduction of the compass, position, destination, and direction at sea were primarily determined by the sighting of landmarks, supplemented with the observation of the position of celestial bodies. On cloudy days, the Vikings may have used cordierite or some other birefringent crystal to determine the sun's direction and elevation from the polarization of daylight; their astronomical knowledge was sufficient to let them use this information to determine their proper heading.[10] For more southerly Europeans unacquainted with this technique, the invention of the compass enabled the determination of heading when the sky was overcast or foggy. This enabled mariners to navigate safely far from land, increasing sea trade, and contributing to the Age of Discovery.

Geomancy and feng shui

Magnetism was originally used, not for navigation, but for geomancy and fortune-telling by the Chinese. The earliest Chinese magnetic compasses were probably not designed for navigation, but rather to order and harmonize their environments and buildings in accordance with the geomantic principles of *feng shui*. These early compasses were made using lodestone, a special form of the mineral magnetite that aligns itself with the Earth's magnetic field.[11]

Based on Krotser and Coe's discovery of an Olmec hematite artifact in Mesoamerica, radiocarbon dated to 1400-1000 BC, astronomer John Carlson has hypothesized that the Olmec might have used the geomagnetic lodestone earlier than 1000 BC for geomancy, a method of divination, which if proven true, predates the Chinese use of magnetism for feng shui by a millennium.[12] Carlson speculates that the Olmecs used similar artifacts as a directional device for astronomical or geomantic purposes but does not suggest navigational usage. The artifact is part of a polished hematite (lodestone) bar with a groove at one end (possibly for sighting). The artifact now consistently points 35.5 degrees west of north, but may have pointed north-south when whole. Carlson's claims have been disputed by other scientific researchers, who have suggested that the artifact is actually a constituent piece of a decorative ornament and not a purposely built compass.[13] [14] Several other hematite or magnetite artifacts have been found at pre-Columbian archaeological sites in Mexico and Guatemala.[15] [16]

Navigational compass

The invention of the navigational compass is credited by scholars to the ancient Chinese, who began using it for navigation sometime between the 9th and 11th century.[14] Europeans and Arabs were first introduced to the compass through nautical contacts during the Chinese Song Dynasty (960–1279).[14] Later the compass appeared in Europe, India, and the Middle East due to the formation of the Mongol Empire which effectly eliminated all previous national barriers within the empire and allowed the safe transfer and transportation of both people and intellectual knowledge across the silk road from China to Europe, the Middle East, and East Africa.

China

Further information: Four Great Inventions, List of Chinese inventions, and History of science and technology in China

There is disagreement as to exactly when the compass was invented. These are noteworthy Chinese literary references in evidence for its antiquity:

Model of a Han Dynasty (206 BC–220 AD) south-indicating ladle or *sinan*. (Historical existence is disputed.)[17]

- The earliest Chinese literature reference to **magnetism** lies in the 4th century BC writings of Wang Xu (): "The lodestone attracts iron."[18] The book also notes that the people of the state of Zheng always knew their position by means of a "south-pointer"; some authors suggest that this refers to early use of the compass.[19]

- The first mention of the **attraction of a needle by a magnet** is a Chinese work composed between 70 and 80 AD (*Lunheng* ch. 47): "A lodestone attracts a needle." This passage of Louen-heng is the first Chinese text concerning the attraction of a needle by a magnet.[20] In 1948, the scholar Wang Chen-Tuo constructed a "compass" in the form of south-indicating spoon on the basis of this text. However, "there is no explicit mention of a magnet in the *Louen-heng*" and that "beforehand it needs to assume some hypotheses to arrive at such a conclusion."[17]

- The earliest reference to a specific magnetic **direction finder** device is recorded in a Song Dynasty book dated to 1040-44. There is a description of an iron "south-pointing fish" floating in a bowl of water, aligning itself to the south. The device is recommended as a means of orientation "in the obscurity of the night." The *Wujing Zongyao* (, "Collection of the Most Important Military Techniques") stated: "When troops encountered gloomy weather or dark nights, and the directions of space could not be distinguished...they made use of the [mechanical] south-pointing carriage, or the south-pointing fish."[21] This was achieved by heating of metal (especially if steel), known today as thermoremanence, and would have been capable of producing a weak state of magnetization.[21] While the Chinese achieved magnetic remanence and induction by this time, a similar discovery was not made in Europe until about 1600, when William Gilbert published his *De Magnete*.[22]

- The first incontestable reference to a **magnetized needle** in Chinese literature appears in 1088.[23] The *Dream Pool Essays*, written by the Song Dynasty polymath scientist Shen Kuo, contained a detailed description of how geomancers magnetized a needle by rubbing its tip with lodestone, and hung the magnetic needle with one single strain of silk with a bit of wax attached to the center of the needle. Shen Kuo pointed out that a needle prepared this way sometimes pointed south, sometimes north.
- The earliest explicit recorded use of a magnetic compass for navigational purposes is found in Zhu Yu's book *Pingzhou Table Talks* (; Pingzhou Ketan) and dates from 1117:[14] *The navigator knows the geography, he watches the stars at night, watches the sun at day; when it is dark and cloudy, he watches the compass.*

Thus, the use of a magnetic compass as a direction finder occurred sometime before 1044, but incontestable evidence for the use of the compass as a navigational device did not appear until 1117.

The typical Chinese navigational compass was in the form of a magnetic needle floating in a bowl of water.[24] According to Needham, the Chinese in the Song Dynasty and continuing Yuan Dynasty did make use of a dry compass, although this type never became as widely used in China as the wet compass.[25] Evidence of this is found in the *Shilin guangji* ("Guide Through the Forest of Affairs"), published in 1325 by Chen Yuanjing, although its compilation had taken place between 1100 and 1250.[25] The dry compass in China was a dry suspension compass, a wooden frame crafted in the shape of a turtle hung upside down by a board, with the lodestone sealed in by wax, and if rotated, the needle at the tail would always point in the northern cardinal direction.[25] Although the European compass-card in box frame and dry pivot needle was adopted in China after its use was taken by Japanese pirates in the 16th century (who had in turn learned of it from Europeans),[26] the Chinese design of the suspended dry compass persisted in use well into the 18th century.[14] [27] However, according to Kreutz there is only a single Chinese reference to a dry-mounted needle (built into a pivoted wooden tortoise) which is dated to between 1150 and 1250, and claims that there is no clear indication that Chinese mariners ever used anything but the floating needle in a bowl until the 16th-century.[24]

The first recorded use of a 48 position mariner's compass on sea navigation was noted in a book titled "The Customs of Cambodia" by Yuan Dynasty diplomat Zhou Daguan, he described his 1296 voyage from Wenzhou to Angkor Thom in detail; when his ship set sail from Wenzhou, the mariner took a needle direction of "ding wei" position, which is equivalent to 22.5 degree SW. After they arrived at Baria, the mariner took "Kun Shen needle", or 52.5 degree SW.[28] Zheng He's Navigation Map, also known as "The Mao Kun Map", contains a large amount of detail "needle records" of Zheng He's travel.[29]

There is a debate over the diffusion of the compass after its first appearance with the Chinese. At present, according to Kreutz, scholarly consensus is that the Chinese invention predates the first European mention by 150 years.[2] However, there are questions over diffusion, because of the apparent failure of the Arabs to function as

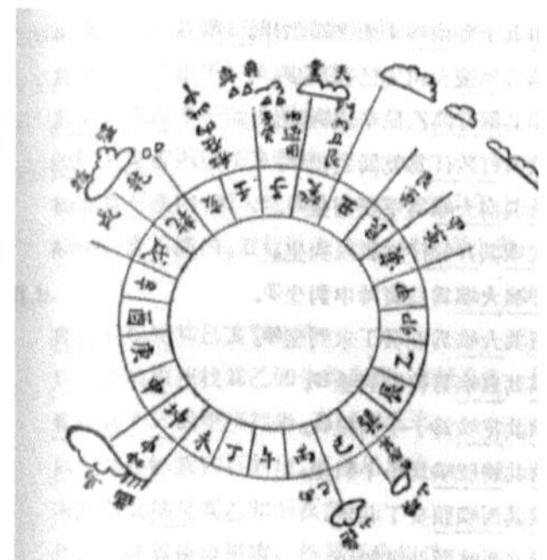
Diagram of a Ming Dynasty mariner's compass

possible intermediaries between East and West because of the earlier recorded appearance of the compass in Europe (1190)[30] than in the Muslim world (1232, 1242, and 1282).[31] [32] The first European mention of a magnetized needle and its use among sailors occurs in Alexander Neckam's *De naturis rerum* (On the Natures of Things), probably written in Paris in 1190.[30] Other evidence for this includes the Arabic word for "Compass" (*al-konbas*), possibly being a derivation of the old Italian word for compass. In the Arab world, the earliest reference comes in *The Book of the Merchants' Treasure*, written by one Baylak al-Kibjaki in Cairo about 1282.[31] Since the author describes having witnessed the use of a compass on a ship trip some forty years earlier, some scholars are inclined to antedate its first appearance accordingly. There is also a slightly earlier non-Mediterranean Muslim reference to an iron fish-like compass in a Persian talebook from 1232.[32]

Medieval Europe

In 1187 Alexander Neckam reported the use of a magnetic compass for the region of the English Channel.[33] In 1269 Petrus Peregrinus of Maricourt described a floating compass for astronomical purposes as well as a dry compass for seafaring, in his well-known Epistola de magnete.[33] In the Mediterranean, the introduction of the compass, at first only known as a magnetized pointer floating in a bowl of water,[34] went hand in hand with improvements in dead reckoning methods, and the development of Portolan charts, leading to more navigation during winter months in the second half of the 13th century.[35] While the practice from ancient times had been to curtail sea travel between October and April, due in part to the lack of dependable clear skies during the Mediterranean winter, the prolongation of the sailing season resulted in a gradual, but sustained increase in shipping movement; by around 1290 the sailing season could start in late January or February, and end in December.[36] The additional few months were of considerable economic importance. For instance, it enabled Venetian convoys to make two round trips a year to the Levant, instead of one.[37]

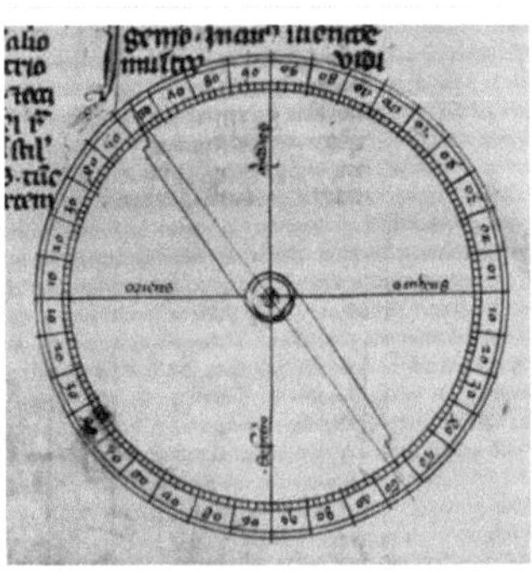

Pivoting compass needle in a 14th century copy of *Epistola de magnete* of Peter Peregrinus (1269)

At the same time, traffic between the Mediterranean and northern Europe also increased, with first evidence of direct commercial voyages from the Mediterranean into the English Channel coming in the closing decades of the 13th century, and one factor may be that the compass made traversal of the Bay of Biscay safer and easier.[38] However, critics like Kreutz feel that it was later in 1410 that anyone really started steering by compass.[39]

At present, according to Kreutz, "barring the discovery of new evidence, it seems clear the first Chinese reference to" the compass "antedates any European mention by roughly 150 years."[2] However, there are questions over diffusion, because of the apparent failure of the Arabs to function as possible intermediaries between East and West because of the earlier recorded appearance of the compass in Europe (1190)[30] than in the Muslim world (1232, 1242, and 1282).[31] [32] This is countered by evidence of the temporal proximity of the Chinese navigational compass (1117) to its first appearance in Europe (1190) and the common shape of the early compass as a magnetized needle floating in a bowl of water.[30]

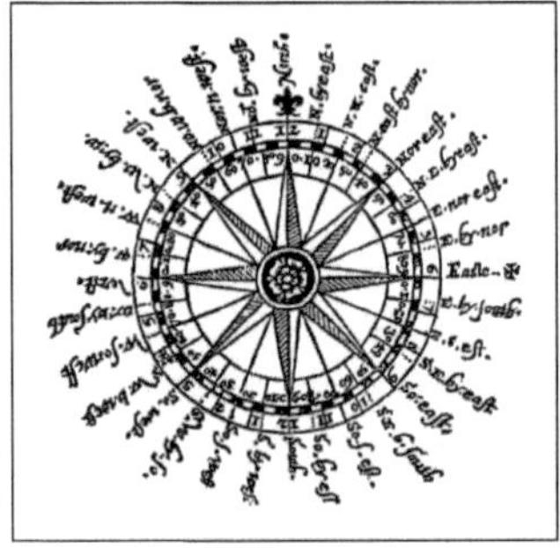

Navigational sailor's compass rose.

Islamic world

The earliest reference to an iron fish-like compass in the Islamic world occurs in a Persian talebook from 1232.[32] This fish shape was from a typical early Chinese design.[14] The earliest Arabic reference to a compass — in the form of magnetic needle in a bowl of water — comes from the Yemeni sultan and astronomer Al-Ashraf in 1282.[31]

He also appears to be the first to make use of the compass for astronomical purposes.[40] Since the author describes having witnessed the use of a compass on a ship trip some forty years earlier, some scholars are inclined to antedate its first appearance in the Arab world accordingly.[32]

In 1300, another Arabic treatise written by the Egyptian astronomer and muezzin Ibn Sim'ūn describes a dry compass for use as a "Qibla (Kabba) indicator" to find the direction to Mecca. Like Peregrinus' compass, however, Ibn Sim'ūn's compass did not feature a compass card.[33] In the 14th century, the Syrian astronomer and timekeeper Ibn al-Shatir (1304–1375) invented a timekeeping device incorporating both a universal sundial and a magnetic compass. He invented it for the purpose of finding the times of Salah prayers.[41] Arab navigators also introduced the 32-point compass rose during this time.[42]

India

The compass was used in India for navigational purposes and was known as the matsya yantra, because of the placement of a metallic fish in a cup of oil.[43]

Medieval Africa

There is evidence that the distribution of the compass from China likely also reached eastern Africa by way of trade through the end of the Silk Road that ended in East African center of trade in Somalia and the Swahili city-state kingdoms. There is evidence that Swahili maritime merchants and sailors acquired the compass at some point and used them for navigation of Swahili versions of dhows.[44]

Later developments

Dry compass

The dry mariner's compass was invented in Europe around 1300. The dry mariner's compass consists of three elements: A freely pivoting needle on a pin enclosed in a little box with a glass cover and a wind rose, whereby "the wind rose or compass card is attached to a magnetized needle in such a manner that when placed on a pivot in a box fastened in line with the keel of the ship the card would turn as the ship changed direction, indicating always what course the ship was on".[3] Later, compasses were often fitted into a gimbal mounting to reduce grounding of the needle or card when used on the pitching and rolling deck of a ship.

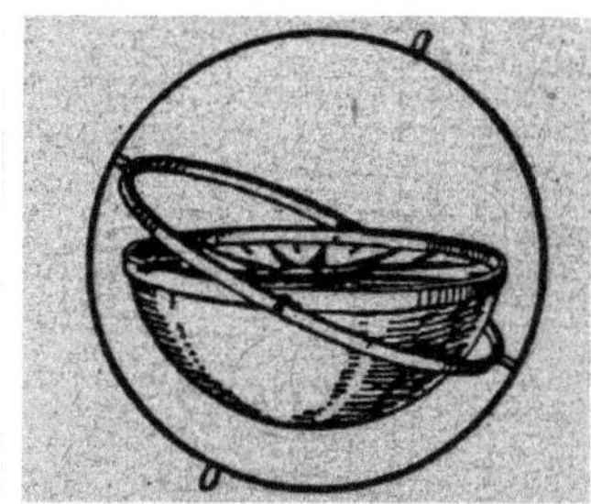

Early modern dry compass suspended by a gimbal (1570)

While pivoting needles in glass boxes had already been described by the French scholar Peter Peregrinus in 1269,[45] and by the Egyptian scholar Ibn Sim'ūn in 1300,[33] traditionally Flavio Gioja (fl. 1302), an Italian pilot from Amalfi, has been credited with perfecting the sailor's compass by suspending its needle over a compass card, thus giving the compass its familiar appearance.[46] Such a compass with the needle attached to a rotating card is also described in a commentary on Dante's *Divine Comedy* from 1380, while an earlier source refers to a portable compass in a box (1318),[47] supporting the notion that the dry compass was known in Europe by then.[24]

Bearing compass

A *bearing compass* is a magnetic compass mounted in such a way that it allows the taking of bearings of objects by aligning them with the lubber line of the bearing compass.[48] A *surveyor's compass* is a specialized compass made to accurately measure heading of landmarks and measure horizontal angles to help with map making. These were already in common use by the early 18th century and are described in the 1728 Cyclopaedia. The bearing compass was steadily reduced in size and weight to increase portability, resulting in a model that could be carried and operated in one hand. In 1885, a patent was granted for a hand compass fitted with a viewing prism and lens that enabled the user to accurately sight the heading of geographical landmarks, thus creating the *prismatic compass*.[49] Another sighting method was by means of a reflective mirror. First patented in 1902, the *Bézard compass* consisted of a field compass with a mirror mounted above it.[50] [51] This arrangement enabled the user to align the compass with an objective while simultaneously viewing its bearing in the mirror.[50] [52]

Bearing compass (18th century).

In 1928, Gunnar Tillander, a Swedish unemployed instrument maker and avid participant in the sport of orienteering, invented a new style of bearing compass. Dissatisfied with existing field compasses, which required a separate protractor in order to take bearings from a map, Tillander decided to incorporate both instruments into a single instrument. It combined a compass with a protractor built into the base His design featured a metal compass capsule containing a magnetic needle with orienting marks mounted into a transparent protractor baseplate with a lubber line (later called a *direction of travel indicator*). By rotating the capsule to align the needle with the orienting marks, the course bearing could be read at the lubber line. Moreover, by aligning the baseplate with a course drawn on a map - ignoring the needle - the compass could also function as a protractor. Tillander took his design to fellow orienteers Björn, Alvid, and Alvar Kjellström, who were selling basic compasses, and the four men modified Tillander's design.[53] In December 1932, the Silva Company was formed with Tillander and the three Kjellström brothers, and the company began manufacturing and selling its Silva orienteering compass to Swedish orienteers, outdoorsmen, and army officers.[54] [55] [56] [53]

Liquid compass

The liquid compass is a design in which the magnetized needle or card is damped by fluid to protect against excessive swing or wobble, improving readability while reducing wear. A rudimentary working model of a liquid compass was introduced by Sir Edmund Halley at a meeting of the Royal Society in 1690.[57] However, as early liquid compasses were fairly cumbersome and heavy, and subject to damage, their main advantage was aboard ship. Protected in a binnacle and normally gimbal-mounted, the liquid inside the compass housing effectively damped shock and vibration, while eliminating excessive swing and grounding of the card caused by the pitch and roll of the vessel. The first liquid mariner's compass believed practicable for limited use was patented by the Englishman Francis Crow in 1813.[58] [59] Liquid-damped marine compasses for ships and small boats were

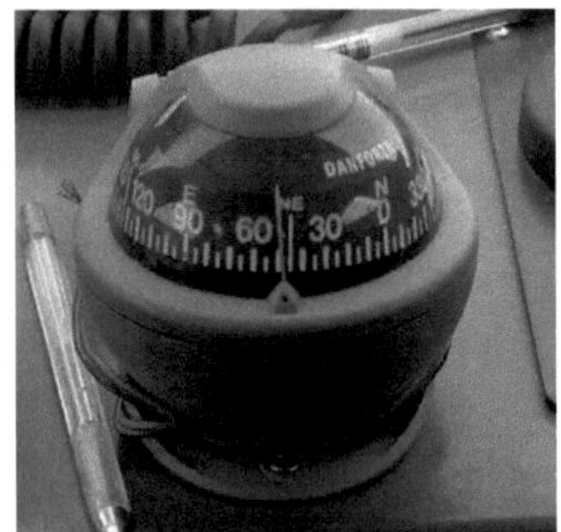
A flush mount, liquid-filled compass on a boat

occasionally used by the British Royal Navy from the 1830s through 1860, but the standard Admiralty compass remained a dry-mount type.[60] In the latter year, the American physicist and inventor Edward Samuel Ritchie patented a greatly improved liquid marine compass that was adopted in revised form for general use by the United States Navy, and later purchased by the Royal Navy as well.[61]

Despite these advances, the liquid compass was not introduced generally into the Royal Navy until 1908. An early version developed by RN Captain Creak proved to be operational under heavy gunfire and seas, but was felt to lack navigational precision compared with the design by Lord Kelvin:

> Captain Creak's first step in the development of the liquid compass was to introduce a "card mounted on a float, with two thin and relatively short needles, fitted with their poles at the scientifically correct angular distances, and with the centre of gravity, centre of buoyancy, and the point of suspension in correct relation to each other...The compass thus designed rectified the defects of the Admiralty Standard Compass...with the additional advantage of considerable steadiness under heavy gunfire and in a seaway... The one defect in the compass as developed by Creak up to 1892 was that "for manoeuvring purposes it was inferior to Lord Kelvin's compass, owing to comparative sluggishness on a large alteration of course through the drag on the card by the liquid in which it floated...[4] [62]

Typical aircraft-mounted magnetic compass

However, with ship and gun sizes continuously increasing, the advantages of the liquid compass over the Kelvin compass became unavoidably apparent to the Admiralty, and after widespread adoption by other navies, the liquid compass was generally adopted by the Royal Navy as well.[4]

Liquid compasses were next adapted for aircraft. In 1909, Captain F.O. Creagh-Osborne, Superintendent of Compasses at the British Admiralty, introduced his *Creagh-Osborne* aircraft compass, which used a mixture of alcohol and distilled water to damp the compass card.[63] [64] After the success of this invention, Capt. Creagh-Osborne adapted his design to a much smaller pocket model[65] for individual use[66] by officers of artillery or infantry, receiving a patent in 1915.[67]

In December 1932, the newly founded Silva Company of Sweden introduced its first baseplate or bearing compass that used a liquid-filled capsule to damp the swing of the magnetized needle.[53] The liquid-damped Silva took only four seconds for its needle to settle in comparison to thirty seconds for the original version.[53]

In 1933 Tuomas Vohlonen, a surveyor by profession, applied for a patent for a unique method of filling and sealing a lightweight celluloid compass housing or capsule with a petroleum distillate to dampen the needle and protect it from shock and wear caused by excessive motion.[68] Introduced in a wrist-mount model in 1936 as the Suunto Oy *Model M-311*, the new capsule design led directly to the lightweight liquid field compasses of today.[68]

History of non-navigational uses

Building orientation

Evidence for the orientation of buildings by the means of a magnetic compass can be found in 12th century Denmark: one fourth of its 570 Romanesque churches are rotated by 5-15 degrees clockwise from true east-west, thus corresponding to the predominant magnetic declination of the time of their construction.[69] Most of these churches were built in the 12th century, indicating a fairly common usage of magnetic compasses in Europe by then.[70]

Mining

The use of a compass as a direction finder underground was pioneered by the Tuscan mining town Massa where floating magnetic needles were employed for determining tunneling and defining the claims of the various mining companies as early as the 13th century.[71] In the second half of the 15th century, the compass became standard equipment for Tyrolian miners. Shortly afterwards the first detailed treatise dealing with the underground use of compasses was published by a German miner Rülein von Calw (1463–1525).[72]

Astronomy

Three astronomical compasses meant for establishing the meridian were described by Peter Peregrinus in 1269 (referring to experiments made before 1248)[73] In the 1300s, an Arabic treatise written by the Egyptian astronomer and muezzin Ibn Sim'ūn describes a dry compass for use as a "Qibla indicator" to find the direction to Mecca. Ibn Sim'ūn's compass, however, did not feature a compass card nor the familiar glass box.[33] In the 14th century, the Syrian astronomer and timekeeper Ibn al-Shatir (1304–1375) invented a timekeeping device incorporating both a universal sundial and a magnetic compass. He invented it for the purpose of finding the times of Salah prayers.[41] Arab navigators also introduced the 32-point compass rose during this time.[42]

Modern compasses

Modern compasses usually use a magnetized needle or dial inside a capsule completely filled with a liquid (lamp oil, mineral oil, white spirits, purified kerosene, or ethyl alcohol is common). While older designs commonly incorporated a flexible rubber diaphragm or airspace inside the capsule to allow for volume changes caused by temperature or altitude, some modern liquid compasses utilize smaller housings and/or flexible capsule materials to accomplish the same result.[74] The liquid inside the capsule serves to dampen the movement of the needle, reducing oscillation time and increasing stability. Key points on the compass, including the north end of the needle are often marked with phosphorescent, photoluminescent, or self-luminous materials[75] to enable the compass to be read at night or in poor light. As the compass fill liquid is noncompressible under pressure, many ordinary liquid-filled compasses will operate accurately underwater to considerable depths.

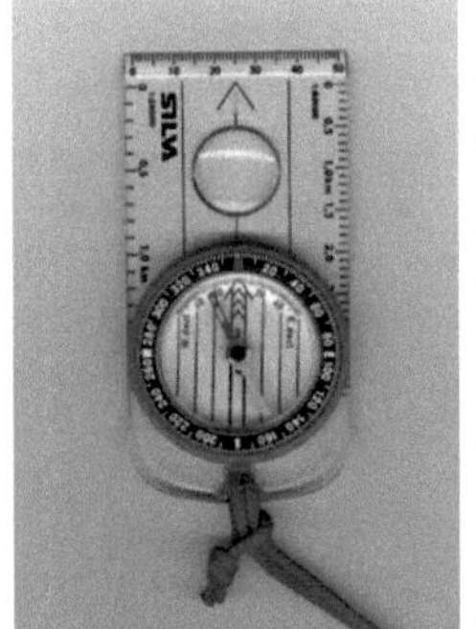
A liquid-filled protractor or orienteering compass with lanyard

Many modern compasses incorporate a baseplate and protractor tool, and are referred to variously as "orienteering", "baseplate", "map compass" or "protractor" designs. This type of compass uses a separate magnetized needle inside a rotating capsule, an orienting "box" or gate for aligning the needle with magnetic north, a transparent base containing map orienting lines, and a bezel (outer dial) marked in degrees or other units of angular measurement.[76] The capsule is mounted in a transparent baseplate containing a *direction-of-travel* (DOT) indicator for use in taking bearings directly from a map.[76]

Liquid filled lensatic compass

Cammenga air filled lensatic compass

Other features found on modern orienteering compasses are map and romer scales for measuring distances and plotting positions on maps, luminous markings on the face or bezels, various sighting mechanisms (mirror, prism, etc.) for taking bearings of distant objects with greater precision, "global" needles for use in differing hemispheres, adjustable declination for obtaining instant true bearings without resort to arithmetic, and devices such as clinometers for measuring gradients.[76] The sport of orienteering has also resulted in the development of models with extremely fast-settling and stable needles for optimal use with a topographic map, a land navigation technique known as *terrain association*.[77]

The military forces of a few nations, notably the United States Army, continue to issue field compasses with magnetized compass dials or cards instead of needles. A magnetic card compass is usually equipped with an optical, lensatic, or prismatic sight, which allows the user to read the bearing or azimuth off the compass card while simultaneously aligning the compass with the objective (see photo). Magnetic card compass designs normally require a separate protractor tool in order to take bearings directly from a map.[76] [78]

The U.S. M-1950 military lensatic compass does not use a liquid-filled capsule as a dampening mechanism, but rather electromagnetic induction to control oscillation of it magnetized card. A "deep-well" design is used to allow the compass to be used globally with a card tilt of up to 8 degrees without impairing accuracy.[79] As induction forces provide less damping than liquid-filled designs, a needle lock is fitted to the compass to reduce wear, operated by the folding action of the rear sight/lens holder. The use of air-filled induction compasses has declined over the years, as they may become inoperative or inaccurate in freezing temperatures or extremely humid environments due to condensation or water ingress.[80]

Some military compasses, like the U.S. M-1950 (Cammenga 3H) military lensatic compass, the Silva 4b *Militaire*, and the Suunto M-5N(T) contain the radioactive material tritium (3H) and a combination of phosphors.[81] The U.S. M-1950 equipped with self-luminous lighting contains 120 mCi (millicuries) of tritium. The purpose of the tritium and phosphors is to provide illumination for the compass, via radioluminescent tritium illumination, which does not require the compass to be "recharged" by sunlight or artificial light.[82]

Mariner's compasses can have two or more gimbaled magnets permanently attached to a compass card. These move freely on a pivot. A *lubber line*, which can be a marking on the compass bowl or a small fixed needle indicates the ship's heading on the compass card. Traditionally the card is divided into thirty-two points (known as *rhumbs*), although modern compasses are marked in degrees rather than cardinal points. The glass-covered box (or bowl) contains a suspended gimbal within a binnacle. This preserves the horizontal position.

Thumb compass

A **thumb compass** is a type of compass commonly used in orienteering, a sport in which map reading and terrain association are paramount. Consequently, most thumb compasses have minimal or no degree markings at all, and are normally used only to orient the map to magnetic north. Thumb compasses are also often transparent so that an orienteer can hold a map in the hand with the compass and see the map through the compass.

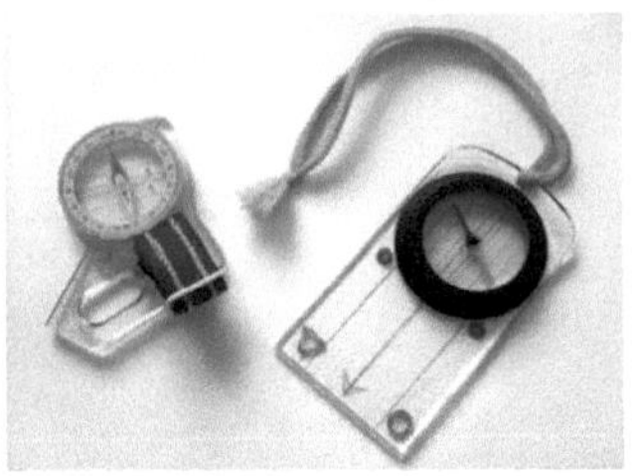

Thumb compass on left

Gyrocompass

A *gyrocompass* is similar to a gyroscope. It is a non-magnetic compass that finds true north by using an (electrically powered) fast-spinning wheel and friction forces in order to exploit the rotation of the Earth. Gyrocompasses are widely used on ships. They have two main advantages over magnetic compasses:

- they find *true north*, i.e., the direction of Earth's rotational axis, as opposed to magnetic north,
- they are not affected by ferromagnetic metal (including iron, steel, cobalt, nickel, and various alloys) in a ship's hull. (No compass is affected by nonferromagnetic metal, although a magnetic compass will be affected by any kind of wires with electric current passing through them.)

Large ships typically rely on a gyrocompass, using the magnetic compass only as a backup. Increasingly, electronic fluxgate compasses are used on smaller vessels. However compasses are still widely in use as they can be small, use simple reliable technology, are comparatively cheap, often easier to use than GPS, require no energy supply, and unlike GPS, are not affected by objects, e.g. trees, that can block the reception of electronic signals.

Solid state compasses

Small compasses found in clocks, mobile phones, and other electronic devices are solid-state compasses, usually built out of two or three magnetic field sensors that provide data for a microprocessor. The correct heading relative to the compass is calculated using trigonometry.

Often, the device is a discrete component which outputs either a digital or analog signal proportional to its orientation. This signal is interpreted by a controller or microprocessor and used either internally, or sent to a display unit. The sensor uses highly calibrated internal electronics to measure the response of the device to the Earth's magnetic field.

GPS receivers using two or more antennae can now achieve 0.5° in heading accuracy (e.g.[83]) and have startup times in seconds rather than hours for gyrocompass systems. Manufactured primarily for maritime applications, they can also detect pitch and roll of ships.

Specialty compasses

Apart from navigational compases, other specialty compasses have also been designed to accommodate specific uses. These include:

A standard Brunton Geo, used commonly by geologists

- Qibla compass, which is used by Muslims to show the direction to Mecca for prayers.
- Optical or prismatic hand-bearing compass, most often used by surveyors, but also by cave explorers, foresters, and geologists. This compasses ordinarily uses a liquid-damped capsule[84] and magnetized floating compass dial with an integral optical (direct or lensatic) or prismatic sight, often fitted with built-in photoluminescent or battery-powered illumination.[76] Using the optical or prism sight, such compasses can be read with extreme accuracy when taking bearings to an object, often to fractions of a degree. Most of these compasses are designed for heavy-duty use, with high-quality needles and jeweled bearings, and many are fitted for tripod mounting for additional accuracy.[76]
- Trough compasses, mounted in a rectangular box whose length was often several times its width, date back several centuries. They were used for land surveying, particularly with plane tables.

Limitations of the magnetic compass

The compass is very stable in areas close to the equator, which is far from "magnetic north". As the compass is moved closer and closer to one of the magnetic poles of the Earth, the compass becomes more sensitive to crossing its magnetic field lines. At some point close to the magnetic pole the compass will not indicate any particular direction but will begin to drift. Also, the needle starts to point up or down when getting closer to the poles, because of the so-called magnetic inclination. Cheap compasses with bad bearings may get stuck because of this and therefore indicate a wrong direction.

All magnetic devices are subject to fields other than Earth's, which is not particularly strong. Local environments may contain mineral deposits and human sources such as MRIs. Vehicles may contain ferrous metals, which may pick up their own fields. Cars may be mostly steel, and render simple compasses useless after time. While ships, submarines, and spacecraft may be built from carefully controlled materials, and later degaussed, drivers rarely take such a step.

A compass is also subject to errors when the compass is accelerated or decelerated in an airplane or automobile. Depending on which of the Earth's hemispheres the compass is located and if the force is acceleration or deceleration the compass will increase the indicated heading or decrease the indicated heading.

Another error of the mechanical compass is turning error. When one turns from a heading of east or west the compass will lag behind the turn or lead ahead of the turn. Magnetometers, and substitutes such as gyrocompasses, are more stable in such situations.

Construction of a compass

Magnetic needle

A magnetic rod is required when constructing a compass. This can be created by aligning an iron or steel rod with Earth's magnetic field and then tempering or striking it. However, this method produces only a weak magnet so other methods are preferred. For example, a magnetised rod can be created by repeatedly rubbing an iron rod with a magnetic lodestone. This magnetised rod (or magnetic needle) is then placed on a low friction surface to allow it to freely pivot to align itself with the magnetic field. It is then labeled so the user can distinguish the north-pointing from the south-pointing end; in modern convention the north end is typically marked in some way, often by being painted red.

Needle-and-bowl device

If a needle is rubbed on a lodestone or other magnet, the needle becomes magnetized. When it is inserted in a cork or piece of wood, and placed in a bowl of water it becomes a compass. Such devices were universally used as compass until the invention of the box-like compass with a 'dry' pivoting needle sometime around 1300.

Points of the compass

Originally, many compasses were marked only as to the direction of magnetic north, or to the four cardinal points (north, south, east, west). Later, these were divided, in China into 24, and in Europe into 32 equally spaced points around the compass card. For a table of the thirty-two points, see compass points.

In the modern era, the 360-degree system took hold. This system is still in use today for civilian navigators. The degree system spaces 360 equidistant points located clockwise around the compass dial. In the 19th century some European nations adopted the "grad" (also called grade or gon) system instead, where a right angle is 100 grads to give a circle of 400 grads. Dividing grads into tenths to give a circle of 4000 decigrades has also been used in armies.

Wrist compass of the Soviet Army with counterclockwise double graduation: 60° (like a watch) and 360°.

Most military forces have adopted the French "millieme" system. This is an approximation of a milli-radian (6283 per circle), in which the compass dial is spaced into 6400 units or "mils" for additional precision when measuring angles, laying artillery, etc. The value to the military is that one angular mil subtends approximately one metre at a distance of one kilometer. Imperial Russia used a system derived by dividing the circumference of a circle into chords of the same length as the radius. Each of these was divided into 100 spaces, giving a circle of 600. The Soviet Union divided these into tenths to give a circle of 6000 units, usually translated as "mils". This system was adopted by the former Warsaw Pact countries (Soviet Union, GDR etc.), often counterclockwise (see picture of wrist compass). This is still in use in Russia.

Compass balancing (magnetic dip)

Because the Earth's magnetic field's inclination and intensity vary at different latitudes, compasses are often balanced during manufacture so that the dial or needle will be level, eliminating needle drag which can give inaccurate readings. Most manufacturers balance their compass needles for one of five zones, ranging from zone 1, covering most of the Northern Hemisphere, to zone 5 covering Australia and the southern oceans. This individual zone balancing prevents excessive dipping of one end of the needle which can cause the compass card to stick and give false readings.

Some compasses feature a special needle balancing system that will accurately indicate magnetic north regardless of the particular magnetic zone. Other magnetic compasses have a small sliding counterweight installed on the needle itself. This sliding counterweight, called a 'rider', can be used for counter balancing the needle against the dip caused by inclination if the compass is taken to a zone with a higher or lower dip.

Compass correction

Like any magnetic device, compasses are affected by nearby ferrous materials, as well as by strong local electromagnetic forces. Compasses used for wilderness land navigation should not be used in proximity to ferrous metal objects or electromagnetic fields (car electrical systems, automobile engines, steel pitons, etc.) as that can affect their accuracy.[76] Compasses are particularly difficult to use accurately in or near trucks, cars or other mechanized vehicles even when corrected for deviation by the use of built-in magnets or other devices. Large amounts of ferrous metal combined with the on-and-off electrical fields caused by the vehicle's ignition and charging systems generally result in significant compass errors.

A binnacle containing a ship's steering compass, with the two iron balls which correct the effects of ferromagnetic materials. This unit is on display in a museum.

At sea, a ship's compass must also be corrected for errors, called deviation, caused by iron and steel in its structure and equipment. The ship is *swung*, that is rotated about a fixed point while its heading is noted by alignment with fixed points on the shore. A compass deviation card is prepared so that the navigator can convert between compass and magnetic headings. The compass can be corrected in three ways. First the lubber line can be adjusted so that it is aligned with the direction in which the ship travels, then the effects of permanent magnets can be corrected for by small magnets fitted within the case of the compass. The effect of ferromagnetic materials in the compass's environment can be corrected by two iron balls mounted on either side of the compass binnacle. The coefficient representing the error in the lubber line, while the ferromagnetic effects and the non-ferromagnetic component.

A similar process is used to calibrate the compass in light general aviation aircraft, with the compass deviation card often mounted permanently just above or below the magnetic compass on the instrument panel. Fluxgate electronic compasses can be calibrated automatically, and can also be programmed with the correct local compass variation so as to indicate the true heading.

Using a compass

A magnetic compass points to magnetic north pole, which is approximately 1,000 miles from the true geographic North Pole. A magnetic compass's user can determine true North by finding the magnetic north and then correcting for variation and deviation. Variation is defined as the angle between the direction of true (geographic) north and the direction of the meridian between the magnetic poles. Variation values for most of the oceans had been calculated and published by 1914.[85] Deviation refers to the response of the compass to local magnetic fields caused by the presence of iron and electric currents; one can partly compensate for these by careful location of the compass and the placement of compensating magnets

Turning the compass scale on the map (D - the local magnetic declination)

under the compass itself. Mariners have long known that these measures do not completely cancel deviation; hence, they performed an additional step by measuring the compass bearing of a landmark with a known magnetic bearing. They then pointed their ship to the next compass point and measured again, graphing their results. In this way, correction tables could be created, which would be consulted when compasses were used when traveling in those locations.

Mariners are concerned about very accurate measurements; however, casual users need not be concerned with differences between magnetic and true North. Except in areas of extreme magnetic declination variance (20 degrees or more), this is enough to protect from walking in a substantially different direction than expected over short distances, provided the terrain is fairly flat and visibility is not impaired. By carefully recording distances (time or paces) and magnetic bearings traveled, one can plot a course and return to one's starting point using the compass alone.[76]

When the needle is aligned with and superimposed over the outlined orienting arrow on the bottom of the capsule, the degree figure on the compass ring at the direction-of-travel (DOT) indicator gives the magnetic bearing to the target (mountain).

Soldier using a prismatic compass to get an azimuth.

Compass navigation in conjunction with a map (*terrain association*) requires a different method. To take a map bearing or *true bearing* (a bearing taken in reference to true, not magnetic north) to a destination with a protractor compass, the edge of the compass is placed on the map so that it connects the current location with the desired destination (some sources recommend physically drawing a line). The orienting lines in the base of the compass dial are then rotated to align with actual or true north by aligning them with a marked line of longitude (or the vertical margin of the map), ignoring the compass needle entirely.[76] The resulting *true bearing* or map bearing may then be read at the degree indicator or direction-of-travel (DOT) line, which may be followed as an *azimuth* (course) to the destination. If a *magnetic* north bearing or *compass bearing* is desired, the compass must be adjusted by the amount of magnetic declination before using the bearing so that both map and compass are in agreement.[76] In the given example, the large mountain in the second photo was selected as the target destination on the map. Some compasses allow the scale to be adjusted to compensate for the local magnetic declination; if adjusted correctly, the compass will give the true bearing instead of the magnetic bearing.

The modern hand-held protractor compass always has an additional direction-of-travel (DOT) arrow or indicator inscribed on the baseplate. To check one's progress along a course or azimuth, or to ensure that the object in view is indeed the destination, a new compass reading may be taken to the target if visible (here, the large mountain). After pointing the DOT arrow on the baseplate at the target, the compass is oriented so that the needle is superimposed over the orienting arrow in the capsule. The resulting bearing indicated is the magnetic bearing to the target. Again, if one is using "true" or map bearings, and the compass does not have preset, pre-adjusted declination, one must additionally add or subtract magnetic declination to convert the *magnetic bearing* into a *true bearing*. The exact value of the magnetic declination is place-dependent and varies over time, though declination is frequently given on the map itself or obtainable on-line from various sites. If the hiker has been following the correct path, the compass' corrected (true) indicated bearing should closely correspond to the true bearing previously obtained from the map.

Compasses are to be laid down on a leveled surface so the needle could point to the magnetic north more accurately, as to that the needle only rests or hangs on a bearing fused to the compass casing, if used at a tilt, the needle might hit the casing on the compass, and hence, not move. This will give a faulty reading . To see if the needle is well leveled, look closely at the needle, and tilt it slightly to see if the needle is swaying side to side freely and the needle not contacting the casing of the compass. If the needle tilts to one direction, tilt the compass slightly and gently to the opposing direction until the compass needle is horizontal, lengthwise. Items to avoid around compasses are magnets of any kind and any electronics. Magnetic fields from electronics can easily disrupt the needle, avoiding it from pointing with the earth's magnetic fields, causing interference. The earth's natural magnetic forces are considerably weak, measuring at 0.5 Gauss and magnetic fields from household electronics can easily exceed it, overpowering the compass needle. Exposure to strong magnets, or magnetic interference can sometimes cause the magnetic poles of the compass needle to differ or even reverse. Avoid iron rich deposits when using a compass, for example, certain rocks which contain magnetic minerals, like Magnetite.This is often indicated by a rock with a surface which is dark and has a metallic luster, not all magnetic mineral bearing rocks have this indication. To see if a rock or an area is causing interference on a compass, get out of the area, and see if the needle on the compass moves. If it does, it means that the area or rock the compass was previously at/on is causing interference and should be avoided.

See also

* Absolute bearing
* Astrocompass
* Beam compass
* Boxing the compass
* Brunton compass
* Coordinates
* Earth Inductor Compass
* Fibre optic gyrocompass
* Fluxgate compass
* Gyrocompass
* Hand compass
* Inertial navigation system
* Magnetic dip
* Marching line
* Magnetic Declination
* Pelorus (instrument)
* Radio compass
* Radio direction finder
* Relative bearing
* Wrist compass

Notes

[1] Merrill, Ronald T.; McElhinny, Michael W. (1983). *The Earth's magnetic field: Its history, origin and planetary perspective* (2nd printing ed.). San Francisco: Academic press. p. 1. ISBN 0-12-491242-7.

[2] Kreutz, p. 367

[3] Lane, p. 615

[4] W. H. Creak: "The History of the Liquid Compass", *The Geographical Journal*, Vol. 56, No. 3 (1920), pp. 238-239

[5] The Earth's magnetic field is approximately that of a tilted dipole. If it were exactly dipolar, the compass would point to the geomagnetic poles, which would be identical to the North Magnetic Pole and South Magnetic Pole; however, it is not, so these poles are not equivalent and the compass only points approximately at the geomagnetic poles.

[6] Seidman, David, and Cleveland, Paul, *The Essential Wilderness Navigator*, Ragged Mountain Press (2001), ISBN 0-07-136110-3, p. 147: Since the magnetic compass is simple, durable, and requires no separate electrical power supply, it remains popular as a primary or secondary navigational aid, especially in remote areas or where power is unavailable.

[7] Serway, Raymond A.; Chris Vuille (2006). *Essentials of college physics* (http://books.google.com/books?id=8n4NCyRgUMEC& pg=PA493&dq=magnet+"north+pole"+earth#v=onepage&q=magnet "north pole" earth&f=false). USA: Cengage Learning. p. 493.

ISBN 0-495-10619-4. .

[8] Emiliani, Cesare (1992). *Planet Earth: Cosmology, Geology, and the Evolution of Life and Environment* (http://books.google.com/
books?id=MfAGpVq8gpQC&pg=PA228&dq=magnet+"north+pole"+earth#v=onepage&q=magnet "north pole" earth&f=false). UK:
Cambridge University Press. p. 228. ISBN 0-521-40949-7. .

[9] Manners, Joy (2000). *Static Fields and Potentials* (http://books.google.com/books?id=vJyqbRPsXYQC&pg=PA148&dq=magnet+
"north+pole"+earth#v=onepage&q=magnet "north pole" earth&f=false). USA: CRC Press. p. 148. ISBN 0-7503-0718-8. .

[10] Gábor Horváth et al (2011). "On the trail of Vikings with polarized skylight". *Philosophical Transactions of the Royal Society B* **366** (1565):
772–782. doi:10.1098/rstb.2010.0194.

[11] "National High Magnetic Field Laboratory: Early Chinese Compass" (http://www.magnet.fsu.edu/education/tutorials/museum/
chinesecompass.html). Florida State University. . Retrieved 2009-02-05.

[12] John B. Carlson, "Lodestone Compass: Chinese or Olmec Primacy? Multidisciplinary Analysis of an Olmec Hematite Artifact from San
Lorenzo, Veracruz, Mexico", *Science*, New Series, Vol. 189, No. 4205 (5 September 1975), pp. 753-760 (1975)

[13] Needham, Joseph; Lu Gwei-Djen (1985). *Trans-Pacific Echoes and Resonances: Listening Once Again*. World Scientific. p. 21.

[14] Temple, Robert (2007). *The genius of China: 3,000 years of science, discovery & invention* (3rd ed.). London: Andre Deutsch. pp. 162–166.
ISBN 978-0-233-00202-6.

[15] A. P. Guimarães, "Mexico and the early history of magnetism", *Revista Mexicana de Fisica*, Vol. 50, pp. 51-53 (2004)

[16] http://www.dartmouth.edu/~izapa/CS-MM-Chap.%203.htm

[17] Li Shu-hua, p. 180

[18] http://www.gutenberg.org/cache/epub/7209/pg7209.html

[19] Needham p. 190

[20] Li Shu-hua, p. 176

[21] Needhamn, p. 252

[22] Temple, p. 156.

[23] Li Shu-hua, p. 182f.

[24] Kreutz, p. 373

[25] Needham p. 255

[26] Needham, p. 289.

[27] Needham, p. 290

[28] Zhou

[29] Ma, Appendix 2

[30] Kreutz, p. 368

[31] Kreutz, p. 369

[32] Kreutz, p. 370

[33] Schmidl, Petra G. (1996–1997). "Two Early Arabic Sources On The Magnetic Compass". *Journal of Arabic and Islamic Studies* **1**: 81–132
http://www.uib.no/jais/v001ht/01-081-132schmidl1.htm#_ftn4

[34] Kreutz, p. 368–369

[35] Lane, p. 606f.

[36] Lane, p. 608

[37] Lane, p. 608 & 610

[38] Lane, p. 608 & 613

[39] Kreutz, p. 372–373

[40] Emilie Savage-Smith (1988), "Gleanings from an Arabist's Workshop: Current Trends in the Study of Medieval Islamic Science and
Medicine", *Isis* **79** (2): 246-266 [263]

[41] (King 1983, pp. 547–8)

[42] G. R. Tibbetts (1973), "Comparisons between Arab and Chinese Navigational Techniques", *Bulletin of the School of Oriental and African
Studies* **36** (1): 97-108 [105-6]

[43] *The American journal of science - Google Books* (http://books.google.com/?id=HUAPAAAAIAAJ&dq="matsya+yantra"+++
compass&q="matsya+yantra"). 1919. . Retrieved 2009-06-30.

[44] The Earth and Its Peoples: A Global History, Volume 2 By Richard Bulliet, Pamela Kyle Crossley, Daniel Headrick, Steven Hirsch, Lyman
John. pg. 381 (http://books.google.com/books?id=MM-0F0fR7E4C&pg=PA381&lpg=PA381&dq=Swahili+sailors+compass&
source=bl&ots=rgtf8_GUKY&sig=msc1KG5HFwqgIj-VC9LyciOisMw&hl=en&ei=fbbPTsCDBaXV0QHc_bkf&sa=X&oi=book_result&
ct=result&resnum=4&ved=0CDAQ6AEwAw#v=onepage&q=Swahili sailors compass&f=false)

[45] Taylor

[46] Lane, p. 616

[47] Kreutz, p. 374

[48] "Hand Bearing Compass" (http://www.westmarine.com/webapp/wcs/stores/servlet/WestAdvisorDisplayView?storeId=30003&
langId=-1&catalogId=10001&advisor=bearing.htm). West Marine. 2004. . Retrieved 2007-12-28.

[49] Frazer, Persifor, *A Convenient Device to be Applied to the Hand Compass*, Proceedings of the American Philosophical Society, Vol. 22, No.
118 (Mar., 1885), p. 216

[50] The Compass Museum, *The Bézard Compass*, Article (http://www.compassmuseum.com/hand/bezard.htm)

[51] Barnes, Scott, Churchill, James, and Jacobson, Cliff, *The Ultimate Guide to Wilderness Navigation*, Globe Pequot Press (2002), ISBN 1-58574-490-5, ISBN 978-1-58574-490-9, p. 27

[52] Barnes, p. 27

[53] Litsky, Frank, *Bjorn Kjellstrom, 84, Orienteer and Inventor of Modern Compass*, Obituaries, The New York Times, 1 September 1995

[54] Seidman, p. 68

[55] Kjellström, Björn, *19th Hole: The Readers Take Over: Orienteering*, Sports Illustrated, 3 March 1969

[56] Silva Sweden AB, *Silva Sweden AB and Silva Production AB Become One Company: History*, Press Release 28 April 2000

[57] Gubbins, David, *Encyclopedia of Geomagnetism and Paleomagnetism*, Springer Press (2007), ISBN 1-4020-3992-1, ISBN 978-1-4020-3992-8, p. 67

[58] Fanning, A.E., *Steady As She Goes: A History of the Compass Department of the Admiralty*, HMSO, Department of the Admiralty (1986), pp. 1-10

[59] Gubbins, p. 67

[60] Fanning, A.E., pp. 1-10

[61] Warner, Deborah, *Compasses and Coils: The Instrument Business of Edward S. Ritchie*, Rittenhouse, Vol. 9, No. 1 (1994), pp. 1-24

[62] Gubbins, p. 67: The use of parallel or multiple needles was by no means a new development; their use in dry-mount marine compasses was pioneered by navigation officers of the Dutch East India Company as early as 1649.

[63] Davis, Sophia, *Raising The Aerocompass In Early Twentieth-century Britain*, British Journal for the History of Science, published online by Cambridge University Press, 15 Jul 2008, pp. 1-22

[64] Colvin, Fred H., *Aircraft Mechanics Handbook: A Collection of Facts and Suggestions from Factory and Flying Field to Assist in Caring for Modern Aircraft*, McGraw-Hill Book Co. Inc. (1918), pp. 347-348

[65] The Compass Museum, Article (http://www.compassmuseum.com/wrist/wrist_1.htm#C-O): Though the *Creagh-Osborne* was offered in a wrist-mount model, it proved too bulky and heavy in this form.

[66] Hughes, Henry A., *Improvements in prismatic compasses with special reference to the Creagh-Osborne patent compass*, Transactions of The Optical Society 16, London: The Optical Society (1915), pp. 17-43: The first liquid-damped compass compact enough for pocket or pouch was the *Creagh-Osborne*, patented in 1915 in Great Britain.

[67] Hughes, Henry A., pp. 17-43

[68] Suunto Oy, *Suunto Company History*, December 2001 Article (http://www.suunto.com/suunto/Worlds/outdoor/main/ outdoor_article_normal.jsp?)

[69] N. Abrahamsen: "Evidence for Church Orientation by Magnetic Compass in Twelfth-Century Denmark", *Archaeometry*, Vol. 32, No. 2 (1992), pp. 293-303 (293)

[70] N. Abrahamsen: "Evidence for Church Orientation by Magnetic Compass in Twelfth-Century Denmark", Archaeometry, Vol. 32, No. 2 (1992), pp. 293-303 (303)

[71] Ludwig and Schmidtchen, p. 62–64

[72] Ludwig and Schmidtchen, p. 64

[73] Taylor, p. 1f.

[74] *Gear Review: Kasper & Richter Alpin Compass*, OceanMountainSky.Com

[75] Nemoto & Co. Ltd., Article (http://www.nemoto.co.jp/en/products/luminova/luminova.html): In addition to ordinary phosphorescent luminous paint (zinc sulfide), brighter photoluminescent coatings of strontium aluminate or isotopes of self-luminous tritium are now being used on modern compasses.

[76] Johnson, G. Mark (2003-03-26). *The Ultimate Desert Handbook*. McGraw-Hill Professional. p. 110. ISBN 0-07-139303-X.

[77] Kjernsmo, Kjetil, '[www.learn-orienteering.org/old/buying.html How to use a Compass], *retrieved 8 April 2012*

[78] U.S. Army, *Map Reading and Land Navigation*, FM 21-26, Headquarters, Dept. of the Army, Washington, D.C. (7 May 1993), ch. 11, pp. 1-3: Any 'floating card' type compass with a straightedge or centerline axis can be used to read a map bearing by orienting the map to magnetic north using a drawn magnetic azimuth, but the process is far simpler with a protractor compass.

[79] ' Article MIL-PRF-10436N (http://landnavigation.org/Documents/Compass Mil Specs.pdf), *rev. 31 October 2003, Washington, D.C.: U.S. Dept. of Defense*

[80] Kearny, Cresson H., *Jungle Snafus...And Remedies*, Oregon Institute Press (1996), ISBN 1-884067-10-7, pp. 164-170: In 1989, one U.S. Army jungle infantry instructor reported that about 20% of the issue lensatic compasses in his company used in a single jungle exercise in Panama were ruined within three weeks by rain and humidity.

[81] Ministry of Defence, *Manual of Map Reading and Land Navigation*, HMSO Army Code 70947 (1988), ISBN 0-11-772611-7, ISBN 978-0-11-772611-6, ch. 8, sec. 26, pp. 6-7; ch. 12, sec. 39, p. 4

[82] "Military Compass" (http://www.orau.org/PTP/collection/radioluminescent/armycompass.htm). Orau.org. . Retrieved 2009-06-30.

[83] "GPS Satellite Compasses" (http://www.psicompany.com/gps-satellite-compass/). Psicompany.com. 2006-08-10. . Retrieved 2009-06-30.

[84] Kramer, Melvin G., U.S. Patent No. 4175333, *Magnetic Compass*, Riverton, Wyoming: The Brunton Company, pub. 27 November 1979: The *Brunton Pocket Transit*, which uses magnetic induction damping, is an exception.

[85] Wright, Monte, Most Probable Position, University Press of Kansas, Lawrence, 1972, p.7

References

- Admiralty, Great Britain (1915) *Admiralty manual of navigation, 1914*, Chapter XXV: "The Magnetic Compass (continued): the analysis and correction of the deviation", London : HMSO, 525 p.
- Aczel, Amir D. (2001) *The Riddle of the Compass: The Invention that Changed the World*, 1st Ed., New York : Harcourt, ISBN 0-15-600753-3
- Carlson, John B. (1975) "Lodestone Compass: Chinese or Olmec Primacy? (http://www.sciencemag.org/cgi/content/abstract/189/4205/753): Multidisciplinary analysis of an Olmec hematite artifact from San Lorenzo, Veracruz, Mexico", *Science*, **189** (4205 : 5 September), p. 753-760, DOI 10.1126/science.189.4205.753
- Gies, Frances and Gies, Joseph (1994) *Cathedral, Forge, and Waterwheel: Technology and Invention in the Middle Age*, New York : HarperCollins, ISBN 0-06-016590-1
- Gubbins, David, *Encyclopedia of Geomagnetism and Paleomagnetism*, Springer Press (2007), ISBN 1-4020-3992-1, ISBN 978-1-4020-3992-8
- Gurney, Alan (2004) *Compass: A Story of Exploration and Innovation*, London : Norton, ISBN 0-393-32713-2
- Johnson, G. Mark, *The Ultimate Desert Handbook*, 1st Ed., Camden, Maine: McGraw-Hill (2003), ISBN 0-07-139303-X
- King, David A. (1983). "The Astronomy of the Mamluks". *Isis* **74** (4): 531–555. doi:10.1086/353360
- Kreutz, Barbara M. (1973) "Mediterranean Contributions to the Medieval Mariner's Compass", *Technology and Culture*, **14** (3: July), p. 367–383
- Lane, Frederic C. (1963) "The Economic Meaning of the Invention of the Compass", *The American Historical Review*, **68** (3: April), p. 605–617
- Li Shu-hua (1954) "Origine de la Boussole 11. Aimant et Boussole", *Isis*, **45** (2: July), p. 175–196
- Ludwig, Karl-Heinz and Schmidtchen, Volker (1997) *Metalle und Macht: 1000 bis 1600*, Propyläen Technikgeschichte, Berlin : Propyläen-Verl., ISBN 3-549-05633-8
- Ma, Huan (1997) *Ying-yai sheng-lan* [The overall survey of the ocean's shores (1433)], Feng, Ch'eng-chün (ed.) and Mills, J.V.G. (transl.), Bangkok : White Lotus Press, ISBN 974-8496-78-3
- Needham, Joseph (1986) *Science and civilisation in China*, Vol. 4: "Physics and physical technology", Pt. 1: "Physics", Taipei: Caves Books, originally publ. by Cambridge University Press (1962), ISBN 0-521-05802-3
- Needham, Joseph and Ronan, Colin A. (1986) *The shorter Science and civilisation in China : an abridgement of Joseph Needham's original text*, Vol. 3, Chapter 1: "Magnetism and Electricity", Cambridge University Press, ISBN 0-521-25272-5
- Seidman, David, and Cleveland, Paul, *The Essential Wilderness Navigator*, Ragged Mountain Press (2001), ISBN 0-07-136110-3
- Taylor, E.G.R. (1951) "The South-Pointing Needle", *Imago Mundi*, **8**, p. 1–7
- Temple, Robert. (1986). *The Genius of China: 3,000 Years of Science, Discovery, and Invention*. With a foreword by Joseph Needham. New York: Simon and Schuster, Inc. ISBN 0-671-62028-2.
- Williams, J.E.D. (1992) *From Sails to Satellites: the origin and development of navigational science*, Oxford University Press, ISBN 0-19-856387-6
- Wright, Monte Duane (1972) *Most Probable Position: A History of Aerial Navigation to 1941*, The University Press of Kansas, Library of Congress Catalog Card Number 72-79318
- Zhou, Daguan (2007) *The customs of Cambodia*, translated into English from the French version by Paul Pelliot of Zhou's Chinese original by J. Gilman d'Arcy Paul, Phnom Penh : Indochina Books, prev publ. by Bangkok : Siam Society (1993), ISBN 974-8298-25-6

External links

- How to Make a Compass (http://www.magnet.fsu.edu/mediacenter/slideshows/compass/index.html) Audio slideshow from the National High Magnetic Field Laboratory
- Science Friday, " *The Riddle of the Compass* (http://www.sciencefriday.com/pages/2002/May/hour2_053102.html)" (interview with Amir Aczel, first broadcast on NPR on May 31, 2002).
- Paul J. Gans, The Medieval Technology Pages: Compass (http://scholar.chem.nyu.edu/tekpages/compass.html)
- The Tides By Sir William Thomson (Lord Kelvin)
- Evening Lecture To The British Association At The Southampton Meeting on Friday, August 25, 1882 (http://zapatopi.net/kelvin/papers/the_tides.html). Refers to compass correction by Fourier series.
- Arrick Robots. Robotics.com Example implementation for digital solid-state compass. *ARobot Digital Compass App Note* (http://www.robotics.com/arobot/compass.html)
- How a tilt sensor works. David Pheifer (http://www.sensorsmag.com/articles/0500/120/main.shtml)
- The Gear Junkie (http://thegearjunkie.com/the-thumb-compass) - review of two orienteering thumb compasses
- The good compass video (http://www.odoo.tv/The-good-Compass.62.0.html?&L=1) - A video about important abilities a compass should have
- COMPASSIPEDIA, the great virtual Compass Museum (http://www.compassmuseum.com/) gives comprehensive information about all sorts of compasses and how to use them.
- Geography fieldwork (http://geographyfieldwork.com/UsingCompass.htm)
- Travel Island (http://www.travel-island.com/travel.outdoor.gears/how.works.compass.maps.html)
- Compass whistles Seven types and subgroups. (http://whistlemuseum.com/2009/05/30/compass-whistles-seven-types--subgroups-a-strauss-2.aspx?ref=rss)

Low-frequency_radio_range

The **low-frequency radio range** (LFR), also known as the **four-course radio range**, **LF/MF four-course radio range**, **A-N radio range**, **Adcock radio range**, or commonly "**the range**", was the main navigation system used by aircraft for instrument flying in the 1930s and 1940s, until the advent of the VHF omnidirectional range (VOR), beginning in the late 1940s. It was used for en route navigation as well as instrument approaches and holds.[1] [2] [3]

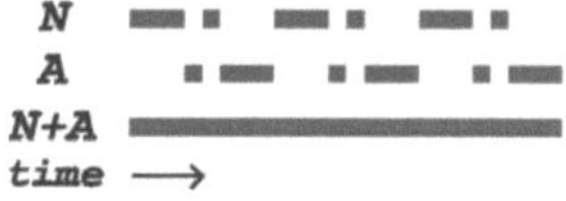

LFR audio signals: *N* stream, *A* stream and combined uniform tone (simulated sounds)

Based on a network of radio towers which transmitted directional radio signals, the LFR defined specific airways in the sky. Pilots navigated the LFR by listening to a stream of automated "A" and "N" Morse codes. For example, they would turn the aircraft to the right when hearing an "N" stream ("dah-dit, dah-dit, ..."), to the left when hearing an "A" stream ("di-dah, di-dah, ..."), and fly straight ahead while hearing a steady tone.[4]

As the VOR system was phased in around the world, the LFR was gradually phased out, mostly disappearing by the 1970s. There are no remaining operational LFR facilities today. At its maximum deployment, there were nearly 400 LFR stations in the U.S. alone.[2]

History

After World War I, aviation began to expand its role into the civilian arena, starting with airmail flights. It soon became apparent that for reliable mail delivery, as well as the passenger flights which were soon to follow, a solution was required for navigation at night and in poor visibility. In the U.S., a network of lighted beacons, similar to maritime lighthouses, was constructed for the airmail pilots. But the beacons were useful mostly at night and in good weather, while in poor visibility conditions they could not be seen. Scientists and engineers realized that a radio based navigation solution would allow pilots to "see" under all flight conditions, and decided a network of directional radio beams was needed.[5]

On September 24, 1929, then-Lieutenant (later General) James H. "Jimmy" Doolittle, U.S. Army, demonstrated the first "blind" flight, performed exclusively by reference to instruments and without outside visibility, and proved that instrument flying was feasible.[6] [7] Doolittle used newly developed gyroscopic instruments—attitude indicator and gyrocompass—to help him maintain his aircraft's attitude and heading, and a specially designed directional radio system to navigate to and from the airport.[8] [9] Doolittle's experimental equipment was purpose-built for his demonstration flights; for instrument flying to become practical, the technology had to be reliable, mass-produced and widely deployed, both on the ground and in the aircraft fleet.[5]

There were two technological approaches for both the ground and air radio navigation components, which were being evaluated during the late 1920s and early 1930s.

On the ground, to obtain directional radio beams with a well defined navigable course, crossed loop antennas were used initially. The first loop-based LFR system was commissioned by the U.S. Commerce Department on June 30, 1928.[2] [5] But the loop antenna design suffered from poor performance, especially at night, and by 1932 the Adcock antenna array, which had superior accuracy, became the preferred solution and replaced the loop antennas. The U.S. Commerce Department's Aeronautics Branch referred to the Adcock solution as the "T-L Antenna" (for "Transmission Line") and did not initially mention Adcock's name.[5] [10]

Jimmy Doolittle demonstrated in 1929 that instrument flying is feasible.

Doolittle's instrument panel

The vibrating reed, developed in the 1920s, was a simple panel mounted instrument with "turn left-right" indicator.

In the air, there were also two competing designs, originating from groups of different backgrounds and needs. The Army Signal Corps, representing military aviators, preferred a solution based on a stream of audio navigation signals, constantly fed into the pilots' ears via a headset. Civilian pilots on the other hand, who were mostly airmail pilots flying cross-country to deliver the mail, felt the audio signals would be annoying and difficult to use over long flights, and preferred a visual solution, with an indicator in the instrument panel.[5]

A visual indicator was developed based on vibrating reeds, which provided a simple panel-mounted "turn left-right" indicator. It was reliable, easy to use and more immune to erroneous signals than the competing audio based system. Pilots who had flown with both aural and visual systems strongly preferred the visual type, according to a published report.[5] [11] The reed-based solution was passed over by the U.S. government, however, and the audio signals became standard for decades to come.[5] [10]

By the 1930s the LFR network of ground-based radio transmitters, coupled with on board AM radio receivers, became a vital part of instrument flying. LFR provided navigational guidance to aircraft for en route operations and approaches, under virtually all weather conditions, helping to make consistent and reliable flight schedules a reality.[4]

LFR remained as the main radio navigation system in the U.S. and other countries until it was gradually replaced by the much-improved VHF-based VOR technology, starting in the late 1940s. The VOR, still used today, includes a visual left-right indicator.[2] [10] [12] [13]

Technology

Ground

The LFR ground component consisted of a network of radio transmission stations which were strategically located around the country, often near larger airports, approximately 200 miles apart. Early LFR stations used crossed loop antennas, but later designs were all based on the Adcock vertical antenna array for improved performance, especially at night.[3] [5]

Each Adcock range station had four 134-foot-tall (**unknown operator: u'strong'** m) antenna towers erected on the corners of a 425 x 425 ft square, with an optional extra tower in the center for voice transmission and homing.[3] [5] [10] The stations emitted directional electromagnetic radiation at 190 to 535 kHz and 1,500 watts, into four quadrants.[1] [14] [15] The radiation of one opposing quadrant pair was

Early LFR station based on crossed loop antennas; later installations used Adcock antennas for improved performance.

modulated (at an audio frequency of 1,020 Hz) with a Morse code for the letter A ($\cdot$ —), and the other pair with the letter N (— $\cdot$).[16] The intersections between the quadrants defined four course lines emanating from the transmitting station, along four compass directions, where the A and N signals were of equal intensity, with their combined Morse codes merging into a steady 1,020 Hz audio tone. These course lines (also called "legs"), where only a tone could be heard, defined the airways.[12]

Adcock range station. The central fifth tower was typically used for voice transmissions.

In addition to the repeating *A* or *N* modulation signal, each transmitting station would also transmit its two-letter Morse code identifier once every thirty seconds for positive identification.[17] The station identification would be sent twice: first on the *N* pair of transmitters, then on the *A*, to ensure coverage in all quadrants.[3] [18] Also, in some installations local weather conditions were periodically broadcast in voice over the range frequency, preempting the navigational signals, but eventually this was done on the central fifth tower.[19] [20]

The LFR was originally accompanied by airway beacons, which were used as a visual backup, especially for night flights.[4] Additional "marker beacons" (low power VHF radio transmitters) were sometimes included as supplementary orientation points.[21]

Air

The airborne radio receivers—initially simple Amplitude Modulation (AM) sets—were tuned to the frequency of the LFR ground transmitters, and the Morse code audio was detected and amplified into speakers, typically in headsets worn by the pilots.[4] The pilots would constantly listen to the audio signal, and attempt to fly the aircraft along the course lines ("flying the beam"), where a uniform tone would be heard. If the signal of a single letter (*A* or *N*) became audibly distinct, the aircraft would be turned as needed so that the modulation of the two letters would overlap again, and the Morse code audio would become a steady tone.[2] The "on course" region, where the *A* and *N* audibly merged, was approximately 3° wide, which translated into a course width of ±2.6 miles when 100 miles away from the station.[4]

Pilots had to verify that they were tuned to the correct range station frequency by comparing its Morse code identifier against the one published on their navigation charts. They would also verify they were flying towards or away from the station, by determining if the signal level (i.e. the audible tone volume) was getting stronger or weaker, respectively.[4]

Chart of Silver Lake LFR, frequency 269 kHz. Aircraft at location 1 would hear: "dah-dit, dah-dit, ...", at 2: "di-dah, di-dah, ...", at 3: a steady tone, and at 4: nothing (Cone of Silence).[22]

Approaches and holds

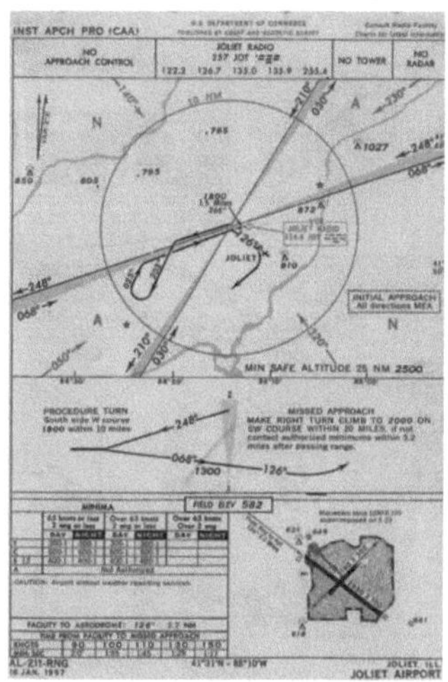
Joliet, IL LFR instrument approach procedure

Final approach segments of LFR instrument approaches were normally flown near the range station, which ensured increased accuracy. When the aircraft was over the station, the audio signal disappeared, since there was no modulation signal directly above the transmitting towers. This quiet zone, called the "Cone of Silence", signified to the pilots that the aircraft was directly overhead the station, serving as a positive ground reference point for the approach procedure.[2] [4]

In a typical LFR instrument approach procedure, final approach would begin over the range station, with a turn to a specific course. The pilot would descend to a specified minimum descent altitude (MDA), and if the airport was not in sight within a specified time (based on ground speed), a missed approach procedure would be initiated. In the depicted Joliet, IL LFR approach procedure, minimum descent altitude could be as low as 300 feet AGL, and required minimum visibility one mile, depending on aircraft type.[21] [23]

The LFR also allowed air traffic control to instruct pilots to enter a holding pattern "on the beam", i.e. on one of the LFR legs, with the holding fix (key turning point) over the LFR station, in the Cone of Silence, or over one of the fan markers. The holds were used either during the en route portion of a flight or as part of the approach procedure near the terminal airport. LFR holds were more accurate than NDB holds, since NDB holding courses are predicated on the accuracy of the on board magnetic compass, whereas the LFR hold was as accurate as the LFR leg, with an approximate course width of $3°$.[3]

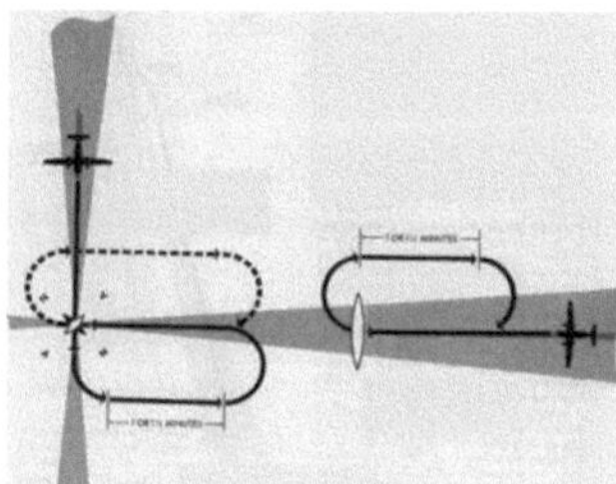
Diagram showing LFR holding patterns.

Non-directional beacons

NDB ground installations are simple single tower antennas.

The LFR required a complex ground installation (four to five antenna towers per station) and careful monitoring of its signals, producing directional radio waves which defined flyable airways, with only a simple AM receiver needed on board the aircraft. From its beginning in the early 1930s, the LFR was augmented with Low Frequency Non-directional beacons (NDBs), which were simple, single-antenna transmitters, whose radio emission pattern was uniform in all directions in the horizontal plane. Coupled with a more complex on-board receiver, called radio direction finder (RDF), used to detect the phase of incoming waves, NDB allowed pilots to determine their azimuth to the transmitting station, and to navigate to or from the station in conjunction with their magnetic compass. Early RDF receivers were costly, bulky and difficult to operate, but the simpler and less expensive ground installation allowed the easy addition of NDB based waypoints and approaches, to supplement the LFR system.[4]

Modern RDF receivers, called "automatic direction finders" are small, low cost and easy to operate, and the NDB remains today as a supplement and backup to VOR and GPS navigation, although it is gradually being phased out.[24] [25]

Limitations

Although the LFR system was used for decades as the main aeronautical navigation method during low visibility and night flying, it had some well known limitations and drawbacks. The course lines, which were a result of a balance between the radiation patterns from different transmitters, would fluctuate depending on weather conditions, vegetation or snow cover near the station, and even the airborne receiver's antenna angle. Under some conditions, the signals from the A quadrant would "skip" into the N quadrant (or vice-versa), causing a false "virtual course" away from any real course line. Also, thunderstorms and other atmospheric disturbances would create electromagnetic interference to disrupt the range signals and produce crackling "static" in the pilots' headsets.[19]

Replacement by VOR

The LFR navigation system required, at a minimum, only a simple AM radio receiver on board the aircraft to accurately navigate the airways under instrument meteorological conditions, and even execute an instrument approach to low minimums.[26] On the downside, however, it had only four course directions per station, was sensitive to atmospheric and other types of interference and aberrations, and required pilots to listen for hours to an annoying monotonous beep, or a faint stream of Morse codes, often embedded in background "static". Its eventual replacement, the VHF band VOR navigation system, had many advantages. The VOR was virtually immune to interference, had

The VHF-based VOR technology superseded the LFR by the 1960s.

360 available course directions, had a visual "on course" display (with no listening needed), and was far easier to use.[13] Consequently, when the VOR system became available in the early 1950s its acceptance was rapid, and within a decade the LFR was mostly phased out. VOR itself is gradually being phased out today in favor of the far superior Global Positioning System (GPS).[10] [25]

Sounds

The following are simulated sounds for the Silver Lake LFR. The range station—located about 10 miles north of Baker, California—would preempt the navigational signals every 30 seconds to transmit its Morse code identifier ("RL"). The station identification would be heard once or twice, possibly with different relative amplitudes, depending on the aircraft location.[3] Pilots would listen to and navigate by these sounds for hours while flying.[2] [4] Actual sounds contained "static", interference and other distortions, not reproduced by the simulation.[4] Adjusting the volume would affect the effective course width.[3] For example, in the simulated sound for "twilight" *A* below, where the aircraft is nearly on the beam but slightly inside the *A* quadrant, a low volume almost obscures the weak *A* sound, whereas a loud one makes it more distinct.

(See Wikipedia:Media help if you encounter problems playing these sound files.)

Notes

[1] Lawrence, Harry (2004). "Airways—from Lighted Beacons to Radio Navigation". *Aviation and the Role of Government*. Kendall Hunt. p. 92. ISBN 0757509444.

[2] "Four-Course Radio Range (Low-Frequency Radio Range (LFR))" (http://www.atcmuseum.org/navigation/nav_sis/nav_sis_lfr.asp). Museum of Air Traffic Control. . Retrieved 2009-07-21.

[3] "The Radio Range". *Instrument Flying - AF Manual 51-37*. Air Training Command, Department of the Air Force. January 20, 1966. pp. 14/1–17.

[4] "On the Beam" (http://www.navfltsm.addr.com/ndb-nav-history.htm). www.navfltsm.addr.com. . Retrieved 2009-07-21.

[5] "BLIND FLYING ON THE BEAM: AERONAUTICAL COMMUNICATION, NAVIGATION AND SURVEILLANCE: ITS ORIGINS AND THE POLITICS OF TECHNOLOGY" (http://ntl.bts.gov/lib/000/700/744/JAT_8-2-7.pdf). Journal of Air Transportation. 2003. .

[6] "Flying Blind: A Brief History of Aviation Advancements, 1918-1930" (http://www.fathom.com/course/10701016/session1.html). Columbia University. . Retrieved 2009-07-24.

[7] "'BLIND' PLANE FLIES 15 MILES AND LANDS; FOG PERIL OVERCOME". *The New York Times*: p. 1. September 25, 1929.

[8] Heppenheimer, T. A. (1995). "Flying Blind" (http://www.americanheritage.com/articles/magazine/it/1995/4/1995_4_54.shtml). *Invention and Technology Magazine* **10** (4). .

[9] Doolittle also used a standard turn and bank instrument which was relatively common by the time of his flight.

[10] "FAA HISTORICAL CHRONOLOGY, 1926-1996" (http://www.faa.gov/about/media/b-chron.pdf). . Retrieved 2009-07-30.

[11] "New Wireless Beacon for Croydon" (http://www.flightglobal.com/pdfarchive/view/1931/1931 - 1255.html). Flight. November 27, 1931. p. 1177. .

[12] Nagaraja (2001). "The LF/MF Four-Course Radio Range". *Elements of Electronic Navigation*. Tata McGraw-Hill. ISBN 007462301X.

[13] *Airport and air traffic control system*. Diane Publishing. 1982. p. 28. ISBN 1428924108.

[14] "Adcock Antenna" (http://web.archive.org/web/20090511075702/http://www.vias.org/radioanteng/rae_01_13_03.html). Virtual Institute of Applied Science. Archived from the original (http://www.vias.org/radioanteng/rae_01_13_03.html) on 2009-05-11. . Retrieved 2009-07-22.

[15] According to international standards, the frequency band below 300kHz is "Low Frequency", and above that "Medium Frequency". Since LFR frequencies "straddled" the dividing line between the two bands, they were technically called "Low Frequency/Medium Frequency (LF/MF) Radio Range" stations.

[16] In the U.S. the quadrant which included the true north radial was designated as *N* (if a course leg was exactly on true north, then the northwest quadrant became *N*); in Canada, *N* was the quadrant which included the 045° true radial.

[17] LFR station identification codes varied between one to three letters.

[18] Since the station identification was transmitted in sequence, first on the *N* and then the *A* antenna pairs, it would be heard by the pilot once or twice, possibly with different relative amplitudes, depending on aircraft location. For example, it would be heard twice when on the beam, and only once when inside a quadrant.

[19] "Flying the Beams". *Popular Mechanics* (Hearst Magazines): 402–404,138A,140A,142A. March 1936. ISSN 0032-4558.

[20] Pilots had to request to stop the weather report if they were using the LFR for an approach.

[21] Thompson, Scott A. (1990). *Flight Check!: The Story Of Faa Flight Inspection*. DIANE Publishing. p. 46. ISBN 0788147285.

[22] Every 30 seconds, Silver Lake station's Morse code identifier, "di-dah-dit di-dah-di-dit" (R-L), would preempt the navigation signals.

[23] U.S. Dept. of Commerce (January 16, 1957). *Joliet Airport Approach Procedure (CAA)*. Coast and Geodetic Survey.

[24] "ADF Basics" (http://www.avweb.com/news/avionics/183233-1.html). September 6, 1998. . Retrieved 2009-07-30.

[25] Clarke, Bill (1998). *Aviator's guide to GPS*. McGraw-Hill Professional. pp. 110–111. ISBN 0070094934.

[26] Basic flight instruments would still be needed.

See also

- SCR-277

References

Further reading

- Oser, Hans J.. *Development of the Visual-Type Airway Radio-Beacon System* (http://permanent.access.gpo. gov/lps1514/nvl.nist.gov/pub/nistpubs/sp958-lide/038-042.pdf). National Institute of Standards and Technology. Retrieved 2009-08-02.

- Diamond, H.; Dunmore, F. W. (September 19, 1930). *A Radio System for Blind Landing of Aircraft in Fog* (http:/ /www.pnas.org/content/16/11/678.full.pdf). Proceedings of the National Academy of Sciences.

- Conway, Erik M. (2006). *Blind landings: low-visibility operations in American aviation, 1918-1958*. JHU Press. ISBN 0801884497.

- "Interview with James H. Doolittle" (http://www.fathom.com/course/10701016/117_transcriptcc.pdf). Columbia University. Retrieved 2009-08-02.

- "Visual-Aural Radio Range" (http://www.airwaysmuseum.com/VAR &Markers Ops Notes 1953.htm). The Airways Museum. Retrieved 2010-05-19.

External links

- Fort Chimo, Quebec LFR approach (http://i100.photobucket.com/albums/m8/Siddley-Hawker/ RNGAPPYVP.jpg)
- The Adcock Range System (http://www.aerofiles.com/adcock-range.html)
- Video clip explaining the LFR (http://www.youtube.com/watch?v=p-VqtNY8vpw)

Navigational_aid

A **navigational aid** (also known as **aid to navigation, ATON,** or **navaid**) is any sort of marker which aids the traveler in navigation; the term is most commonly used to refer to nautical or aviation travel. Common types of such aids include lighthouses, buoys, fog signals, and day beacons.

According to the glossary of terms in the United States Coast Guard Light list, an Aid to Navigation is any device external to a vessel or aircraft specifically intended to assist navigators in determining their position or safe course, or to warn them of dangers or obstructions to navigation.

Lateral markers

Red ATONs always have even numbers, and green ATONs have odd numbers. Under the IALA B standard used in North and South America, when you are going to sea, the red ATON is on your left, and the green on your right. Under the IALA A standard used in Europe, Africa and most of Asia, the colors are reversed.

A lighthouse is an easily recognized aid to navigation.

In the IALA B system, the red ATONs are on your right when you return from sea (Red Right Returning) and the green on your left. Red daybeacons are triangles and green daybeacons are squares. All of these ATONs are Lateral Markers that mark traffic channels and where it is safe to travel.

Non-lateral markers

There are also non-lateral markers that give information other than the edges of safe waters. Most are white with orange markings and black lettering. They are used to give direction and information, warn of hazards and destructions, mark controlled areas, and mark off-limits areas. These ATONs do not mark traffic channels.

On non-lateral markers, there are some shapes that show certain things:

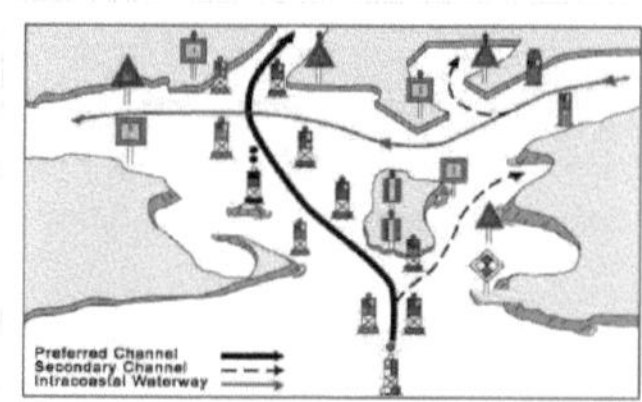

Diagram showing lateral, non-lateral, and special purpose markers as seen on a nautical chart.

- Squares - show information, including places to find food, supplies, and repairs. They sometimes show directions.

- Diamonds - warn about dangers like rocks, construction, dams, or stumps.

- Circles - mark a controlled area such as no wake, idles speed, speed limit, or ski zone.

- Crossed diamonds - show areas off limits to all boats, like swimming areas and dams.

Junction buoys

There are some special colored ATONs. When there are red and green horizontal stripes, you are at the junction of two channels. The ATONs indicate the primary channel. If the green is on the top, the preferred channel is to the right. If the red is on top, the preferred channel is to the left. The light matches the top stripe color. These ATONs are sometimes called "junction buoys."

AIS ATONs

ATONs are often integrated with AIS, e.g. a lighthouse can be equipped with an AIS transitter. Sometimes it is impractical to equip the ATON with an AIS transponder; in this case an AIS shore station can be assigned to transmit AIS messages on behalf of the ATON. This is known as a *synthetic ATON*.

In other cases, such as marking a wreck until a physical buoy can be deployed, a so-called *virtual ATON* is created: A shore-based AIS system is configured to transmit AIS messages indicating the existence of an ATON at a specified location.

Lead marks and lights

Lead marks (as in "leading a ship into a safe place") and lights are fixed markers that are laterally displaced to allow a mariner to navigate a fixed channel along the preferred route. They are also known as "channel markers."[1] They can normally be used coming into and out of the channel. When lit, they are also usable at night. Customarily, the upper mark is up-hill from the lower (forward) mark. The mariner will know the geometry of the marks/lights from the navigational chart and can understand that when "open" (not one above the other) the ship needs to be navigated to "close" the marks (so one is above the other) and be in the preferred line of the channel.

In some cases, the lead marks/lights are provided by lasers, as in the laser channel under the Tasman Bridge on the Derwent River at Hobart, Tasmania.

Triangle shaped lead marks with lights.

Further reading

- United States Coast Guard. *Aids to Navigation*, (Washington, DC: U. S. Government Printing Office, 1945).
- Scott T. Price. "U. S. Coast Guard Aids to Navigation: A Historical Bibliography" [2]. United States Coast Guard Historian's Office.
- UK Department for Transport. *UK Government Strategy for AIS.* [3]
- IALA. *IALA Standard A-126: On the Use of the Automatic Identification System (AIS) in Marine Aids to Navigation Service.* [4]

Notes

[1] Silk, Robert (March 10, 2010). "Channel marker proposal upsets anglers" (http://keysnews.com/node/21481). *keysnews.com* (Key West, Florida: The Citizen). . Retrieved December 2, 2010.

[2] http://www.uscg.mil/History/weblighthouses/h_lhbib.asp

[3] http://www.dft.gov.uk/pgr/shippingports/ports/modern/ais/ukgovernmentstrategyforais?page=2

[4] http://site.ialathree.org/pages/publications/publicationsessaip2.php?LeTypePub=1

See also

large buoy in storage, Homer, Alaska

- USCG aids to navigation boat
- Buoy
- Daymark
- Distance Measuring Equipment (DME)
- Foghorn
- Global Positioning System (GPS)
- Instrument Landing System (ILS)
- landmark
- Lighthouse
- LORAN
- Non-Directional Beacon (NDB)
- Racon
- Radio navigation
- Range light
- Sea mark
- Tactical Air Navigation (TACAN)
- VHF Omni-directional Range (VOR)
- International Regulations for Preventing Collisions at Sea

External links

- Trevor Diamond's Aviation Navaid Gallery. (http://www.trevord.com/navaids)
- Terry Pepper, Seeing the Light. (http://www.terrypepper.com/Lights/index.htm)
- Aids to Navigation in the Gulf of Gdansk (http://ais.strefa.pl)

1946_Australian_National_Airways_DC-3_crash

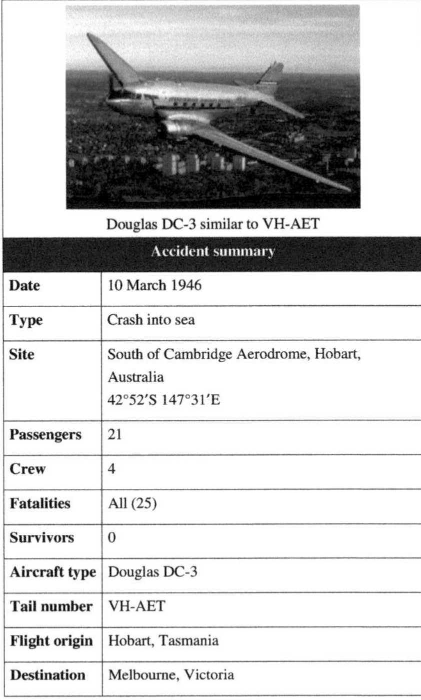

Douglas DC-3 similar to VH-AET

Accident summary	
Date	10 March 1946
Type	Crash into sea
Site	South of Cambridge Aerodrome, Hobart, Australia 42°52′S 147°31′E
Passengers	21
Crew	4
Fatalities	All (25)
Survivors	0
Aircraft type	Douglas DC-3
Tail number	VH-AET
Flight origin	Hobart, Tasmania
Destination	Melbourne, Victoria

On Sunday 10 March 1946 a Douglas DC-3 aircraft departed from Hobart, Australia for a flight to Melbourne. The aircraft crashed into the sea with both engines operating less than 2 minutes after takeoff. All twenty-five people on board the aircraft died. It was Australia's worst civil aviation accident.[1] [2]

An investigation panel was promptly established to investigate the accident. The panel was unable to conclusively establish the cause but it decided the most likely cause was that the automatic pilot was inadvertently engaged shortly after takeoff while the gyroscope was caged. The Department of Civil Aviation took action to ensure that operation of the automatic pilot on-off control on Douglas DC-3 aircraft was made distinctive from operation of any other control in the cockpit, and that instructions were issued impressing on pilots that gyroscopes should be un-caged prior to takeoff.

An inquiry chaired by a Supreme Court judge closely examined three different theories but found there was insufficient evidence to determine any one of them as the cause. This inquiry discovered that the captain of the aircraft was diabetic and had kept it secret from both his employer and the Department of Civil Aviation. The judge considered the captain's diabetes and self-administration of insulin probably contributed significantly to the accident but he stopped short of making this his official conclusion.

In his report, the judge recommended modification of the lever actuating the automatic pilot. The inquiry uncovered four irregularities in the regulation of civil aviation in Australia and the judge made four recommendations to deal with these irregularities.

The flight

The Australian National Airways aircraft registered VH-AET arrived at Cambridge aerodrome at 8:15 pm local time, about four hours late. The return flight to Essendon Airport was scheduled to depart at 4:50 pm, but did not do so until 8:50 pm.[3]

On board were 21 passengers, 3 pilots and an air hostess.[3] Douglas DC-3 (and C-47) aircraft were normally crewed by two pilots but on 10 March the cockpit of VH-AET was occupied by a third person, a supernumerary pilot who was making his first flights with the airline.[4] [5] [6] [7] The weight of the aircraft was about 900 pounds (408 kg) below the maximum authorised weight.[8] [9] The takeoff was into a light southerly wind towards Frederick Henry Bay and the sea. Observers at the aerodrome reported that the takeoff was normal, and both engines were operating perfectly.[3]

Witnesses in the vicinity of Seven-Mile Beach[10] estimated that the aircraft reached a height of a little above 400 ft (122 m)[11] before turning left slightly and descending steeply.[12] The aircraft cleared the land and crashed into Frederick Henry Bay about 300 yards (275 m)[13] beyond the water's edge and a mile (1.6 km) from the western end of Seven-Mile Beach.[3] [14] After takeoff it flew for less than 2 minutes[14] [15] and covered a distance of only 2.9 nautical miles (5.4 km).[16]

Recovery

Wreckage

Rear fuselage inverted on Seven-Mile Beach. The fin and rudder have been folded down parallel to the tailplane.[17]

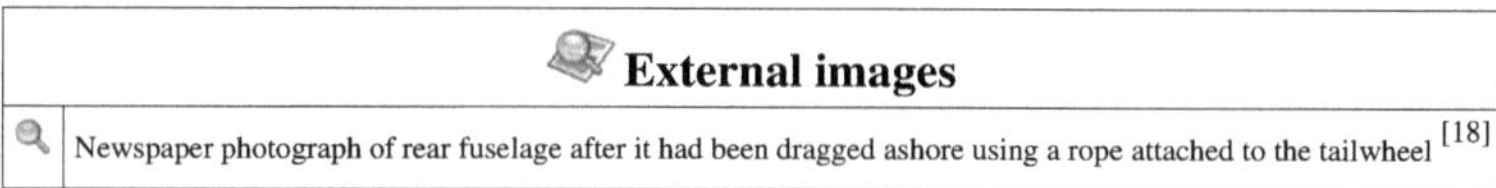

External images

Newspaper photograph of rear fuselage after it had been dragged ashore using a rope attached to the tailwheel [18]

On learning of the crash at nearby Seven-Mile Beach employees of Australian National Airways raced from Cambridge aerodrome to lend assistance. About 11:15 pm the rear fuselage came to the surface a short distance off-shore.[19] [20] Donald Butler,[21] one of the employees, feared the air hostess might still be trapped in her seat in the rear of the fuselage. He took a length of rope, swam out to the floating piece of structure, attached the rope to the tailwheel and then swam back to the beach. A motor lorry was used to drag the rear fuselage ashore but there was no-one inside.[] [] [20] The right tailplane, elevator and trim tab were almost undamaged. The elevator trim tab was still set appropriately for a shallow climb after take-off.[12]

Diver Glen Thorne[22] found pieces of wreckage scattered over a wide area of the sea bed. The aircraft had disintegrated and there were few recognisable pieces of structure.[23] Key parts of the wreckage were eventually recovered from the sea bed by Thorne working from a diving pontoon towed to the site by Royal Australian Navy ship *HMAS Huon*.[24] [] [] The wreckage was in about 18 ft (5.5 m) of water.[25] [26]

Bodies

About 4 hours after the crash the mutilated body of a woman was washed onto Seven-Mile Beach. Fifteen minutes later the body of a man was washed ashore. It was later identified as the body of the captain. At intervals until 6:30 am another 5 bodies were washed ashore.[] The next day, another 3 bodies were recovered.[26] The bodies were badly mutilated and either naked or clad only in vestiges of underclothing, indicating the severity of the impact with the water.[5] [] One body was missing a leg.[] A head, severed from its body, was recovered in the vicinity.[27]

The bodies of 21 of the 25 people on board were eventually recovered from the beaches around Frederick Henry Bay. The remaining 4 bodies were never found.[28] [] [29] One body was found on the beach at Sandford, about 5 miles from the site of the crash.[30] The body of the supernumerary pilot was not identified until 19 days after the accident.[31]

Two years after the accident, a human thigh-bone was found on Seven-Mile Beach. Police believed the bone came from one of the bodies never recovered.[32] [33]

Investigation

The Director-General of Civil Aviation promptly established a panel to investigate the accident.[34] The panel was chaired by John Watkins, Superintendent of Airworthiness & Aeronautical Engineering.[35] [] [36] After examining the wreckage recovered from the sea bed the panel was satisfied both engines were operating at high power at the time of impact and there was no pre-existing structural or mechanical defect that would explain the crash.[12]

The panel arranged for the witnesses at Seven Mile Beach to observe a series of flights by a de Havilland Dragonfly[37] taking off from Cambridge aerodrome, and to identify which flight best represented what they saw on the night of the crash. This exercise established that VH-AET reached a maximum height of about 425 feet (130 m).[] The panel members were satisfied that, after take-off, VH-AET achieved a normal climb of about 325 feet per minute (1.6 m/s) and a gradient of about 1¾% before suddenly commencing a descent of about 17½%. Calculations showed that the aircraft's descent reached about 4,000 feet per minute (20 m/s) and its speed about 160 mph (257 km/hour) before it struck the water.[12]

The panel contemplated 25[38] possible causes of the accident.[39] In its interim report to the Director-General on 20 March 1946[40] the panel reduced these possible causes to 3:[41]

Automatic pilot

The control box for the automatic pilot was recovered from the wreckage. The control knob for its gyroscope[42] was in the caged position and the *CAGED* warning flag was in view even though the control box carried a placard stating the gyro must be uncaged before takeoff. The automatic pilot's speed valve unit was also recovered; it showed that the three valves were open in positions indicating the unit was operating at the time of the crash. The investigation panel proposed that the automatic pilot might have been engaged at a height of about 400 ft (120 m) causing the aircraft to descend swiftly into the sea. Engagement could have occurred inadvertently when one of the pilots intended to select fuel cross-feed ON.[43] [40] [] [44] [45] The operating levers for fuel cross-feed and automatic pilot were the same shape, the same height above the cockpit floor, and about 12 inches (305 mm) apart on the cockpit control console.[]

No-one on the ground at Cambridge aerodrome saw the pilot who occupied the co-pilot seat prior to takeoff. The panel proposed that the captain may have asked the supernumerary pilot to occupy the co-pilot seat during the flight to Essendon Airport. The supernumerary pilot was on his third flight with the company and had no prior experience flying the C-47 or DC-3.[46] The panel considered that if the supernumerary pilot had occupied the co-pilot seat, and if the captain had called for fuel cross-feed to be turned ON to deal with a fuel pressure problem in one engine, it was possible the supernumerary pilot's lack of familiarity with the DC-3 cockpit caused him to inadvertently engage the automatic pilot.[47] [6]

The panel used a DC-3 aircraft with a cockpit identical to that of VH-AET to carry out four flight tests with the gyroscope caged. When the automatic pilot was engaged the control wheel moved forward so strongly it was torn from the pilot's hands and he could only regain control by use of both hands and significant force. Initial tests showed that when the pilot was unprepared, up to 600 ft (183 m) could be lost before he identified the problem and disengaged the automatic pilot. On the fourth test the pilot restricted the height loss to 300 ft (91 m).[] The panel considered the most likely explanation of the accident was inadvertent engagement of the automatic pilot with the gyroscope caged.[6] [39]

Others

- Bird strike - A fortnight after the accident the mutilated body of a large bird was found on Seven-Mile Beach. It was considered the bird, which had a wingspan of about 6 ft (1.8 m), may have struck the aircraft cockpit, distracting or incapacitating the pilots.[48] [49] Alternatively, the bird could have struck the aircraft's pitot tube, damaging it and causing inaccuracy in the airspeed indicator.[7]
- Pilot illness - The investigation panel considered the pilot may have collapsed suddenly.[41] [50] [] The panel used a DC-3 aircraft to carry out tests in which one of the pilots simulated falling forward onto the aircraft controls. They found it difficult to fall forward sufficiently to exert a significant force on the elevator control, and the other pilot had no difficulty keeping the aircraft in a climbing attitude.[]

Report

The investigation panel's report to the Director-General made recommendations including:

1. operation of the automatic pilot on-off control on Australian-registered DC-3 aircraft should be made distinctive from operation of any other control in the cockpit.[51]
2. instructions should be issued impressing on pilots that gyroscopes should be un-caged prior to takeoff.
3. the practice of using passenger-carrying flights to give experience to pilots under instruction should be reviewed urgently.[36]

Inquiry

On 24 April 1946 the Minister for Civil Aviation, Arthur Drakeford, appointed Mr Justice Simpson of the Supreme Court of the Australian Capital Territory to conduct an inquiry into the accident.[52] Counsel assisting the inquiry was to be Henry Winneke.[53] [39]

Justice Simpson examined the evidence in detail, including the evidence put forward in support of the 3 most likely causes identified by the investigation panel. He eventually found there was insufficient evidence to consider any of the theories proved.[54] [55] Justice Simpson's report of the findings of his inquiry was made public by the Minister on 11 June 1946.[56] Simpson said he was satisfied the accident was not caused by failure of any part of the aircraft's structure, its engines or its controls; or failure to remove any of the flight-control clamps prior to takeoff.[57] [58] [59]

Automatic pilot

John Watkins, chairman of the investigation panel, told the inquiry the evidence supporting the theory that inadvertent engagement of the automatic pilot caused the accident was that its control box was recovered from the wreckage and its gyroscope was still caged.[40] The speed valve unit was also recovered and it indicated the automatic pilot was operating at the time of impact. Inadvertent engagement of the automatic pilot with the gyroscope caged could explain a sudden descent by a DC-3.[] Watkins also said the panel's report to the Director-General made several recommendations and they were already being actioned. These included one recommendation that operation of the automatic pilot on-off control on Australian-registered DC-3 aircraft should be made distinctive from operation of any other control in the cockpit.[36]

The Flight Superintendent of Australian National Airways, Captain P.T.L. Taylor, told the inquiry he did not believe the accident could have been caused by inadvertent engagement of the automatic pilot. He said if that happened, the pilot could dis-engage it before losing 50 ft (15 m) in altitude.[60] [7]

The Chief Technical Adviser of Australian National Airways, Thomas Lawrence,[61] told the inquiry he did not think there was any evidence to indicate the automatic pilot was engaged at the time of the accident. He thought the crash was the result of a combination of factors.[62] [63] [64]

Bird-strike

Michael Sharland,[65] honorary ornithologist to the Tasmanian Museum, told the Inquiry he had been shown the mutilated remains of a dead bird and had identified it as a gannet, a fishing bird known to dive on its prey from heights from 50 ft to 500 ft. He was unable to say how the bird had died but said its injuries suggested it had been in a collision with a heavy, fast-moving body.[11] [49] Captain P.T.L. Taylor said he thought a bird-strike on the aircraft's pitot tube could have caused the crash.[7]

Justice Simpson rejected the bird-strike theory, saying the descent of the aircraft was caused by forward-movement of the control column in the cockpit. He was unable to state what caused this movement of the control column.[54] [55] [58]

Medical

The inquiry discovered that the chief-pilot of the aircraft, Captain Thomas Spence,[66] was diabetic and had been discharged from the RAAF in September 1941 as medically unfit. In early 1942 he applied for a commercial pilot licence but did not declare his diabetes.[] [] [67] In a medical examination for the purpose of obtaining the licence, and at all subsequent examinations, Spence had shown no sign of diabetes. A specialist in diabetes[68] told the Inquiry it was almost impossible to detect diabetes in a person who wished to withhold it.[]

The inquiry also received evidence that a friend[69] had asked Spence about his diabetes in relation to his employment as a pilot. Spence had asked his friend to keep it quiet so his employment would not be jeopardised.[39] [70] [71]

The senior route captain for Australian National Airways, Captain Douglas Way,[72] told the inquiry he was unaware Spence was treating himself with insulin. Captain Way said he knew Spence had been discharged from the RAAF as medically unfit but Spence had told him it was a minor complaint discovered when he was in Canada and when he returned to Australia he found himself cured.[62] [63] [64]

At a medical examination in October 1943 for renewal of his commercial pilot licence Spence told the examiner he had been in hospital with influenza and a carbuncle. The examiner did not inquire further on the matter. Investigations with Brisbane Hospital for the purpose of informing Justice Simpson revealed that Spence's hospitalisation was for diabetic pre-coma.[67] Justice Simpson agreed that Spence had misled many people.[73]

Spence was scheduled to have a medical examination on 11 March and it was considered plausible that he may have taken extra insulin in order to prepare himself for the examination.[73] An overdose of insulin, or irregular doses, can distort the senses and cause the muscles to be unco-ordinated.[] [74]

Counsel assisting the Inquiry, Henry Winneke, advocated that the cause of the accident was Spence's diabetes.[73] [75] Justice Simpson was critical of the Director-General's investigation panel for considering 25[76] possible causes of the accident but failing to consider that the Department's negligence in licensing a diabetic pilot might have been the root cause of the accident.[39] After the inquiry had received all available information related to Spence's diabetes, Justice Simpson said the insulin reaction of the pilot might have had a considerable bearing on the accident.[70]

In Justice Simpson's report to the Governor-General he wrote that he could see much to support the theory that the most likely cause was Spence's actions in the cockpit while he was adversely affected by insulin. However, in his report he didn't determine that the accident had been caused by Spence's medical condition because there was insufficient evidence to completely prove the theory.[54] [55] [58]

Irregularities

During the court of inquiry Justice Simpson became aware of four irregularities and listed them in his report.

1. When Thomas Spence had applied for a commercial pilot licence the RAAF Director-General of Medical Services had been acting as assessor for the Department of Civil Aviation. The Director-General had neglected to check Spence's RAAF medical history.
2. A medical examiner had failed to check Spence's statement that his recent hospitalisation was due to influenza. The truth was that Spence had been suffering a serious diabetic condition.
3. VH-AET was approved to carry 24 persons. On 10 March 1946 the aircraft took off with 25 persons on board - a full complement of 21 passengers, an air hostess and 3 pilots instead of the usual 2.[77]
4. No flare path had been lit to illuminate the runway for takeoff and any unexpected return to the aerodrome for landing.[78]

Recommendations

Justice Simpson made five recommendations:

1. The practice of allowing pilots under instruction to gain experience in the cockpit of aircraft carrying passengers should be reviewed.
2. The levers that actuate the automatic pilot and the fuel cross-feed should be modified so they have a different appearance.
3. A regulation should be made to prohibit taking off at night without a flare path or other runway lighting system; and it should be compulsory for the flare path to remain lit until there is no longer a possibility the aircraft might return to the airport to land.
4. Ground engineers completing daily inspections should keep copies of their inspection records.
5. Medical examinations for issue or renewal of pilot licences should be made by doctors chosen, and paid for, by the Department of Civil Aviation.[58] [79]

Coronial inquest

The Tasmania Coroner, Mr Sorell, investigated the deaths of the 21 people whose bodies had been recovered.[] [29] He determined that the causes of their deaths were multiple fractures and injuries but he was unable to say how or in what manner they met their deaths.[14]

Flight crew

The captain was Thomas Spence, aged 30. He had about 3,500 hours flying experience and had been a captain of Douglas DC-3 aircraft for a year. He joined Australian National Airways in June 1942.[12]

The co-pilot was David Collum, aged 21. He had about 1,400 hours flying experience, mostly with Australian National Airways.[12]

The supernumerary pilot was Austin Gibson, aged 37. He had about 2,500 hours flying experience in the RAAF; half of this as a flying instructor. He had over 1,000 hours in command of twin-engine Anson, Oxford, Hudson and Beaufort aircraft but no experience on the Douglas C-47 or DC-3.[12]

Aircraft

The aircraft was constructed in 1942 as a Douglas C-47-DL transport aircraft with a Douglas serial number 6013. It was assigned the US military serial number *41-18652* and in 1943 was delivered to the US Army Air Force in Brisbane. In November 1944, it was sold to the Commonwealth of Australia.[44] Twelve C-47s were purchased by the Commonwealth of Australia and hired out under charter to aviation companies, six to Australian National Airways.[80] [81]

The aircraft was registered VH-AET by the Commonwealth and hired out to Australian National Airways on 20 December 1944.[44] [80] Australian National Airways converted it to the civil aircraft configuration about a year before the crash.[82] [83] VH-AET flew for 7,477 hours.[44]

See also

- 1948 Lutana crash - Douglas DC-3 operated by ANA
- 1949 MacRobertson Miller Aviation DC-3 crash - Accident in Perth, Western Australia
- List of disasters in Australia by death toll
- List of accidents and incidents involving commercial aircraft
- List of accidents and incidents involving airliners by location

Notes

[1] It remained the worst until the crash of the Douglas DC-4 *Amana* in 1950. The Mercury - 28 June 1950 (http://trove.nla.gov.au/ndp/del/
     article/26711871#pstart1904725) Retrieved 2011-09-22 Prior to 10 March 1946 Australia's worst civil aviation accident was the Douglas
     DC-2 *Kyeema* which crashed near Melbourne on 25 October 1938 killing all 18 people on board.

[2] The Canberra Times - 12 March 1946, p.2 (http://trove.nla.gov.au/ndp/del/article/2667710) Retrieved 2011-09-15

[3] The Mercury - 11 March 1946 (http://trove.nla.gov.au/ndp/del/article/26174682) Retrieved 2011-09-14

[4] The cockpit of VH-AET was not equipped with a jump seat or lap strap for a third occupant.

[5] Sydney Morning Herald - 12 March 1946 (http://trove.nla.gov.au/ndp/del/article/17971801) Retrieved 2011-09-21

[6] Job 1992, p. 50

[7] The Mercury - 3 May 1946, p.9 (http://trove.nla.gov.au/ndp/del/article/26205912) Retrieved 2011-10-01

[8] The Certificate of Airworthiness for VH-AET authorised a maximum weight of 25,900 lb (11 748 kg). An alternative source says VH-AET
     was 880 lb (400 kg) below the maximum weight.

[9] The West Australian - 3 May 1946, p.10 (http://trove.nla.gov.au/ndp/del/article/50339201) Retrieved 2011-09-27

[10] Seven-Mile Beach is the northern boundary of Frederick Henry Bay.

[11] The Advocate - 2 May 1946, p.5 (http://trove.nla.gov.au/ndp/del/article/68973739#pstart5812154) Retrieved 2011-09-25

[12] Job 1992, p. 48

[13] Estimates of the distance from the shore to the wreckage vary between 200 yards (183 m) and 400 yards (366 m). The Canberra Times - 14
      March 1946, p.5 (http://trove.nla.gov.au/ndp/del/article/2668039) Retrieved 2011-09-24

[14] The Mercury - 26 July 1946, p.9 (http://trove.nla.gov.au/ndp/del/article/26197475) Retrieved 2011-09-15

[15] The Advocate - 1 May 1946, p.5 (http://trove.nla.gov.au/ndp/del/article/68973571#pstart5812146) Retrieved 2011-09-25

[16] The position of the wreckage of VH-AET was approximately:
      42°52′S 147°31′E

[17] Job 1992, p. 45

[18] http://trove.nla.gov.au/ndp/del/page/1667767

[19] Newspapers published on 12 March stated the rear fuselage came to the surface only 45 minutes after the crash. The Mercury - 12 March
      1946 (http://trove.nla.gov.au/ndp/del/article/26174889) Retrieved 2011-09-24 The Argus - 12 March 1946, p.1 (http://trove.nla.gov.
      au/ndp/del/article/22233459) Retrieved 2011-09-14

[20] Job 1992, p. 42

[21] Donald James Butler was a maintenance engineer employed at Cambridge aerodrome by Australian National Airways. The Mercury - 23
      July 1946, p.11 (http://trove.nla.gov.au/ndp/del/article/26188867) Retrieved 2011-09-20 Butler was able to confirm that Spence did not
      require any maintenance carried out on the aircraft. Prior to takeoff Butler observed Spence occupying the Captain's seat in the cockpit.

[22] Glen Armstrong Thorne was a diver employed by the Hobart Bridge Company. The Mercury - 25 July 1946, p.8 (http://trove.nla.gov.au/
      ndp/del/article/26201264) Retrieved 2011-09-22 The Mercury - 15 March 1946, p.4 (http://trove.nla.gov.au/ndp/del/article/26175643)
      Retrieved 2011-09-19

[23] The Mercury - 14 March 1946, p.13 (http://trove.nla.gov.au/ndp/del/article/26175459) Retrieved 2011-09-22

[24] *HMAS Huon* was a 40 ft wooden naval launch.Job 1992, p. 47

[25] Newspaper reports of the water depth vary between 15 ft and 18 ft.

[26] The Canberra Times - 13 March 1946 (http://trove.nla.gov.au/ndp/del/article/2667791) Retrieved 2011-09-19

[27] The Argus - 13 March 1946, p.3 (http://trove.nla.gov.au/ndp/del/article/22233779) Retrieved 2011-09-22

[28] The four bodies never found were those of the co-pilot (David John Collum), two male passengers and a female passenger. Some newspaper reports spelled the co-pilot's surname as Colburn. The Mercury - 21 March 1946, p.5 (http://trove.nla.gov.au/ndp/del/article/26176778) Retrieved 2011-10-01 Courier-Mail - 13 March 1946, p.3 (http://trove.nla.gov.au/ndp/del/article/50284540) Retrieved 2011-10-01

[29] The Mercury - 29 June 1946, p.4 (http://trove.nla.gov.au/ndp/del/article/26202228) Retrieved 2011-09-19

[30] The Mercury - 25 March 1946, p.7 (http://trove.nla.gov.au/ndp/del/article/26177428) Retrieved 2011-09-21

[31] Adelaide Advertiser - 30 March 1946, p.11 (http://trove.nla.gov.au/ndp/del/article/48698429#pstart2614757) Retrieved 2011-09-22

[32] The Mercury - 19 March 1948, p.4 (http://trove.nla.gov.au/ndp/del/article/26452904#pstart1889886) Retrieved 2011-09-26

[33] Sydney Morning Herald - 19 March 1948, p.4 (http://trove.nla.gov.au/ndp/del/article/18064884) Retrieved 2011-09-19

[34] The Director-General of Civil Aviation was Daniel McVey.Drinkwater, Derek, *Sir Daniel McVey* (http://adb.anu.edu.au/biography/ mcvey-sir-daniel-11029), Australian Dictionary of Biography. Retrieved 2011-12-17 In May 1946 McVey resigned as Director-General and Captain Edgar Charles JohnstonWalker, J.D., *Edgar Charles Johnston* (http://adb.anu.edu.au/biography/johnston-edgar-charles-12701), Australian Dictionary of Biography. Retrieved 2011-12-17 began acting as Director-General. The investigation panel consisted of A. Affleck, H. Duke, W. Ellis, and J. Watkins.

[35] At the time of the inquiry, John Leslie Watkins was acting Superintendent of Airworthiness and Aeronautical Engineering in the Department of Civil Aviation. The Mercury - 12 March 1946, p.4 (http://trove.nla.gov.au/ndp/del/article/26174965) Retrieved 2012-03-20

[36] The Argus - 4 May 1946, p.3 (http://trove.nla.gov.au/ndp/del/article/22255777) Retrieved 2011-09-21

[37] VH-UXS was based at Cambridge aerodrome carrying out aerial survey work for the Tasmanian Government.

[38] The number of possible causes was also given as 22 or 23.

[39] The Advocate - 16 May 1946, p.5 (http://trove.nla.gov.au/ndp/del/article/68975796) Retrieved 2011-09-24

[40] The Mercury - 15 May 1946, p.12 (http://trove.nla.gov.au/ndp/del/article/26187027) Retrieved 2011-09-24

[41] The Mercury - 25 July 1946, p.8 (http://trove.nla.gov.au/ndp/del/article/26201264) Retrieved 2011-09-24

[42] A gyroscope is the core of an aircraft's automatic pilot and attitude indicator. While the rotor is accelerating to operational speed the gyroscope should be caged (locked) to prevent damage. It should be un-caged prior to takeoff.

[43] Most operators of DC-3 and C-47 aircraft carried out the takeoff with fuel cross-feed selected ON, but Australian National Airways carried out the takeoff with cross-feed selected OFF. This variation in procedure was the outcome of a near-fatal accident in 1940 when VH-USY *Bungana*, one of the company's DC-2 aircraft, suffered an engine fire in which the fire continued to be fed fuel from the cross-feed system.Job 1992, p. 49

[44] Aussieairliners (http://www.aussieairliners.org/dc-3/vh-aet/vhaet.html) Retrieved 2011-09-14

[45] Aviation Safety Network (http://aviation-safety.net/database/record.php?id=19460310-0) Retrieved 2011-09-14

[46] Flt.Lt. Austin Clement Gibson had been discharged from the RAAF a week before the accident. He had served as a flying instructor for 5 years and had flown twin-engine Anson, Oxford, Hudson and Beaufort aircraft.Job 1992, p. 48

[47] The captain would have been likely to call for fuel cross-feed to be turned ON if the fuel supply pressure in one engine became inadequate, causing a red warning lamp to illuminate in the cockpit. One witness, Michael Reynolds of Seven Mile Beach, reported hearing an unevenness in the sound of the engines as the aircraft approached. The Mercury - 1 May 1946, p.10 (http://trove.nla.gov.au/ndp/del/article/ 26186670) Retrieved 2012-02-25 This unevenness could have been misfiring due to inadequate fuel supply pressure in one engine.

[48] The Mercury - 1 May 1946 (http://trove.nla.gov.au/ndp/del/article/26186599) Retrieved 2011-09-16

[49] The Mercury - 2 May 1946, p.10 (http://trove.nla.gov.au/ndp/del/article/26184438) Retrieved 2011-09-16

[50] The Advocate - 25 July 1946, p.2 (http://trove.nla.gov.au/ndp/del/article/69048190) Retrieved 2011-09-22

[51] This modification was made mandatory in Australia and remains mandatory vide Airworthiness Directive AD/DC3/9 *Automatic Pilot Control Handle - Modification* Airworthiness Directive DC3 No. 9 (http://www.casa.gov.au/ADFiles/over/dc3/DC3-009.pdf) Retrieved 2012-02-28

[52] The Argus - 25 April 1946 (http://trove.nla.gov.au/ndp/del/article/22242897) Retrieved 2011-09-23

[53] Sir Henry Arthur Winneke was knighted in 1957 and later became Chief Justice of the Supreme Court of Victoria (1964) and Governor of Victoria (1974.)

[54] The Mercury - 12 June 1946, p.1 (http://trove.nla.gov.au/ndp/del/article/26197963) Retrieved 2011-09-15

[55] The Advocate - 12 June 1946, p.5 (http://trove.nla.gov.au/ndp/del/article/68980005#pstart5812436) Retrieved 2011-09-26

[56] Sydney Morning Herald - 12 June 1946, p.3 (http://trove.nla.gov.au/ndp/del/article/17987357) Retrieved 2011-09-24

[57] Allen Lauder Hume (engineer-in-charge) and Donald James Butler (maintenance engineer), both employed at Cambridge aerodrome by Australian National Airways, advised the inquiry that all clamps had been removed from the aircraft's control surfaces before takeoff, undercarriage pins had been removed, and the engines were operating normally.

[58] The Argus - 12 June 1946, p.3 (http://trove.nla.gov.au/ndp/del/article/22251072) Retrieved 2011-09-16

[59] The Canberra Times - 12 June 1946, p.4 (http://trove.nla.gov.au/ndp/del/article/2686329) Retrieved 2011-09-15

[60] Accident Investigator and aviation writer Macarthur Job wrote: "Dr E G Coppel, KC, appeared for ANA. As Dr Coppel saw it, the Panel's assumption, integral to the events they had postulated, that Supernumerary Pilot Gibson was occupying the copilot's seat at the time of the accident, reflected adversely on his client, ANA. The evidence which Dr Coppel now drew on heavily to refute the Panel's theory was that Captain Spence had been a secret diabetic whose condition had gone undetected by medical examiners since he applied for a civil pilot's

licence in 1942."Job 1992, p. 51

[61] Thomas Fulton Coleman Lawrence (1915 - 2003), aeronautical engineer. In 1965, Lawrence became Chief Superintendent of the Aeronautical Research Laboratories. Vale Thomas Fulton Coleman Lawrence - p.4 or page 6 of 27. (http://www.brindabellaskiclub.org.au/i/docs/2003_june.pdf) Retrieved 2011-09-29

[62] The Advocate - 17 May, p.5 (http://trove.nla.gov.au/ndp/del/article/68976021) Retrieved 2011-09-27

[63] The Mercury - 17 May 1946, p.4 (http://trove.nla.gov.au/ndp/del/article/26187958) Retrieved 2011-09-19

[64] Sydney Morning Herald - 17 May 1946, p.5 (http://trove.nla.gov.au/ndp/del/article/29765256) Retrieved 2011-09-20

[65] Michael Stanley Sharland was the author of *Tasmanian Birds*.

[66] Thomas William Spence was born in 1916. In February 1940 he passed a medical examination admitting him to flying training in the RAAF. He completed his flying course in Canada but was then found to be diabetic and unfit for flying. He was repatriated to Australia in April 1941 and discharged from the RAAF in September 1941. In March 1942 he passed a medical examination for initial issue of a commercial pilot licence. The Morning Bulletin - 12 March 1946, p.4 (http://trove.nla.gov.au/ndp/del/article/56393926) Retrieved 2011-09-20 The Advocate - 4 May 1946, p.5 (http://trove.nla.gov.au/ndp/del/article/68974015) Retrieved 2011-09-26

[67] The Mercury - 4 May 1946 (http://trove.nla.gov.au/ndp/del/article/26202761#pstart1876657) Retrieved 2011-09-24

[68] Dr E. Downie of Collins St, Melbourne.

[69] The friend was identified as Mr Whitty. The day after the accident Whitty informed an officer in the Department of Civil Aviation, of Spence's diabetes. The officer sent a memorandum conveying Whitty's information but, apart from calling for Spence's RAAF history file, the investigation panel took no further action. John Watkins said the investigation panel took no further action because it believed Spence's diabetes had no bearing on the accident.

[70] The Argus - 16 May 1946, p.6 (http://trove.nla.gov.au/ndp/del/article/22249498#reloadOnBack) Retrieved 2011-09-24

[71] The Canberra Times - 16 May 1946, p.3 (http://trove.nla.gov.au/ndp/del/article/2681929#pstart687625) Retrieved 2011-09-24

[72] Captain Douglas Rowntree Way.

[73] The Mercury - 18 May 1946, p.2 (http://trove.nla.gov.au/ndp/del/article/26205748) Retrieved 2011-09-24

[74] The Canberra Times - 4 May 1946 (http://trove.nla.gov.au/ndp/del/article/2679907) Retrieved 2011-09-16

[75] The Argus - 18 May 1946, p.8 (http://trove.nla.gov.au/ndp/del/article/22258882) Retrieved 2011-10-02

[76] The number of possible causes was also given as 22 or 23.

[77] Evidence was given that when a temporary seat was installed in the cockpit for a supernumerary pilot, a document could be issued to authorise an extra person being carried.

[78] The Mercury - 12 June 1946 (http://trove.nla.gov.au/ndp/del/article/26197968) Retrieved 2011-09-20

[79] The Mercury - 12 June 1946, p.2 (http://trove.nla.gov.au/ndp/del/article/26198033) Retrieved 2011-09-19

[80] The Argus - 13 March 1946, p.3 (http://trove.nla.gov.au/ndp/del/article/22233784#pstart1667792) Retrieved 2011-09-22

[81] The Mercury - 13 March 1946 (http://trove.nla.gov.au/ndp/del/article/26175097) Retrieved 2011-09-21

[82] The Argus - 12 March 1946, p.3 (http://trove.nla.gov.au/ndp/del/article/22233500) Retrieved 2011-09-18

[83] The Mercury - 1 May 1946, p.10 (http://trove.nla.gov.au/ndp/del/article/26186678) Retrieved 2011-09-19

References

Bibliography

- Job, Macarthur (1992). *Air Crash Vol. 2, Chapter 3*. Aerospace Publications Pty. Ltd. Fyshwick, Australia. pp. 200. ISBN 1-875671-01-3

Article Sources and Contributors

1948_Lutana_crash *Source*: http://en.wikipedia.org/w/index.php?title=1948_Lutana_crash *Contributors*: Dolphin51, Hugo999, Jeff79, Mdnavman, Modal Jig, MoondyneAWB, Pajimbo, Ssbohio, 2 anonymous edits

Douglas_DC-3 *Source*: http://en.wikipedia.org/w/index.php?title=Douglas_DC-3 *Contributors*: 790, Absolute adam, Adry1991, Agateller, Ahoerstemeier, Ahunt, Akradecki, Alan Canon, Alaniaris, Albamhandae, Andrew Gray, Andros 1337, AniRaptor2001, Anomalocaris, AnonMoos, Ansgar Walk, Antheii, Anthony Appleyard, Armistej, Arpingstone, Ballista, BilCat, Binksternet, BjKa, Bobblewik, Bongwarrior, Bort1974, Brecchie, Brutaldeluxe, Bzuk, CambridgeBayWeather, Catsmeat, Chrislk02, Christian List, Clawed, Cliffb, Clipper471, Colin Douglas Howell, Corker1, Cottingham, Crwesq@gmail.com, Cryonic07, DARTH SIDIOUS 2, Dabarkey, Dan Dassow, Darkroom, Darz Mol, David Newton, DavidWBrooks, Dbenbenn, Decora, Deekaygee, Degen Earthfast, Dehk, Denniss, Dirk P Broer, Djmckee1, DocWatson42, Dolphin51, Doontass, Edward, Elf-friend, Elipongo, Emt147, Engineman, Ericg, Erzahler, Flygirl-usa, Flyingd, Fnlayson, Frankwit01, GCarty, Galoubet, Gb6491, Gene Nygaard, Gfoley4, Graham87, GregorB, Greyengine5, Groyolo, Guanaco, Gunnai, Hammersbach, Henrickson, Holyjoe722, Humbaband, Hunter1084, IanOsgood, Impi, Intersofia, Iolaire, Irrgang, Isaac Rabinovitch, Jackehammond, Jeremiestrother, Jeronimo, Jivecat, Jno, JohnOwens, Joseph Solis in Australia, Joshbaumgartner, Karl Dickman, Ketiltrout, Kipio, Knowledgekid87, L1011dal, LOGBOOK, LWF, Ldfifty, Leandrod, Liftarn, Lightmouse, Llevy, Logan, Luna Santin, M@sk, MBK004, Magus732, Mark Sublette, Markle the WikiGnome, MattTM, Maxcook, Mbuday, Mdnavman, Meggar, Meiskam, Michael Hardy, Michael Zimmermann, MilborneOne, Mjroots, MoRsE, Mojave350, Moshe Constantine Hassan Al-Silverburg, Mr.choppers, Msteen, Mukkakukaku, Mwarren us, N328KF, NiD.29, Nick Moss, Noisy, NorthnBound, Nuance 4, Omicronpersei8, Onorem, Ooziewangle, Pankkake, Paterakis, Patrick, PatrickDunfordNZ, Pearle, Petebutt, Petri Krohn, Philip Trueman, Pilatichhunter, Piledhigheranddeeper, Plugwash, Producercunningham, Puddhe, QFSE Media, RG2, Ray Van De Walker, Rcbutcher, Rchamberlain, Reedy, Rhopkins8, Rich Farmbrough, Rimovm, Ringwayobserver, Rlandmann, RobbWiki, Robertson-Glasgow, Rodeime, RottweilerCS, Russavia, RuthAS, Sadharan, Sam Hocevar, Sannse, ScottJ, Sekicho, Septegram, Seth ze, Shadowjams, Skysmurf, Sonett72, Southerndc3, Sp33dyphil, Sporran, Suradnik13, Sylvain Mielot, T1n0, TSRL, Tangentier, Tannin, Technopat, Techspert, Teddks, Teh wonkey bat, Template namespace initialisation script, The Bushranger, Themattmeista, TimothyMN, Timsdad, TomKat222, Tombright, Tommy2010, Towpilot, Trevor MacInnis, Túrelio, Userboy87, Utcursch, Uvaphdman, Vgy7ujm, ViperPilot, Wavelength, Whhalbert, White Star Line Fan, WideArc, Wilee, Wimt, Wordyuk, Wsk, XChaos, YSSYguy, Yummifruitbat, Yuriybrisk, Zaidpjd, Zazpot, Zenstone, ZeroOne, 263 anonymous edits

Australian_National_Airways *Source*: http://en.wikipedia.org/w/index.php?title=Australian_National_Airways *Contributors*: Airtimes, Alan U. Kennington, Ardfern, CarolGray, Choster, Colonies Chris, CommonsDelinker, DI2000, Dolphin51, Dupz, ELECTRA, Eavirtue, Edward, Enyavar, Felix Portier, FiggyBee, Fred Niven, Gittinsj, Glen Dillon, Good Olfactory, Graham87, HenryLi, Hossen27, Hsan22, Iamnotolly, Kevin Ryde, Lester, Melesse, Mjager, Mjroots, Msmonksilver, Navarine, Pajimbo, Pearle, Porturology, Prolog, R'n'B, Russavia, RuthAS, Sardanaphalus, Sendervictorius, Shirt58, Stemonitis, Stevenghetto257, Tabletop, Thewikimonster, Vox3000, Waggers, Willi-willi, Wongm, Woohookitty, YSSYguy, 16 anonymous edits

National_aviation_authority *Source*: http://en.wikipedia.org/w/index.php?title=National_aviation_authority *Contributors*: 2829VC, Ahunt, Arsenikk, Ccgrimm, Chris the speller, CutOffTies, Debresser, Deltaker, Dtom, EoGuy, Felix505, Fnlayson, Funandtrvl, Grafen, Malcolma, Mauls, Mopskatze, Neurolysis, Nomi887, Parler Vous, Rich Farmbrough, Rrekn, Rsrikanth05, Stevebow, Wavelength, WhisperToMe, XLerate, 15 anonymous edits

Supreme_Court_of_the_Australian_Capital_Territory *Source*: http://en.wikipedia.org/w/index.php?title=Supreme_Court_of_the_Australian_Capital_Territory *Contributors*: Adz, Assize, Bidgee, Cahk, Chrism, Clarkk, Cme35, Debresser, Docu, Gaius Cornelius, Good Olfactory, M.O.X, Mathew5000, Miracle Pen, MoondyneAWB, Muzi, Shell Kinney, Sss333, Station1, Stephen Bain, That Guy, From That Show!, Vanished user 5zariu3jisj0j4irj, Xtra, 2 anonymous edits

Compass *Source*: http://en.wikipedia.org/w/index.php?title=Compass *Contributors*: (jarbarf), ***Ria777, .:Ajvol:., 16@r, 49oxen, 4twenty42o, A3RO, A930913, ABF, AGK, AHMartin, Aaron McDaid, Acroterion, Addshore, AdjustShift, Adrian.benko, Aesopos, Afluent Rider, Agamemnus, Ahoerstemeier, Ahunt, Aitias, Ajraddatz, Alansohn, Alexf, Alfio, Allstarecho, Altenmann, Altes2009, Amplitude101, AnakngAraw, Anarchyboy, Ancheta Wis, Anclation, Andrew2536, Anpersonalaccount, Antandrus, Anthony Appleyard, Anthony Nolan O'Nymous, Archanamiya, ArielGold, Armando, Arpingstone, Arsonal, Artobest, Aruton, Arx Fortis, Aspenocean, AssegaiAli, Audriusa, Avenged Eightfold, BRPXQZME, Backslash Forwardslash, Badamsambu, Badly Bradley, Balthazarduju, Bauple58, Beach drifter, BenWilliamson, Benjwong, BernardZ, Betacommand, Bfigura's puppy, Big Brother 1984, Bigdottawa, Billlion, Blaxthos, BlueDevil, BobKawanaka, Bobblewik, Bobo192, Bongwarrior, Boris Barowski, Brandon, Brekdanser, Brian0918, Brockert, Bryan Derksen, Bsadowski1, Bucketsofg, Byron Vickers, CWii, Cacophony, Caiqian, Caltas, Calvin Lourdes He, CambridgeBayWeather, CanOfWorms, CanadianLinuxUser, Canderson7, Canyouhearmenow, Cbramble, Cfailde, CharlieEchoTango, Chetvorno, Chmod007, Chris 73, Chris55, Chrishmt0423, Chrislk02, Ciaccona, ClamDip, Clayhalliwell, Cmh, Cold Season, Cometstyles, Comte0, Cool Blue, Coolpatj, Corpx, Courcelles, Ctjf83, CyberDragon777, Cyfal, D.D.Jacko, DARTH SIDIOUS 2, DGG, DJBullfish, DMacks, DOwenWilliams, DVD R W, Damicatz, Dancter, DanielDeibler, Dante Alighieri, Davewild, Dawn Bard, Dbam, Dcljr, Dellant, Denelson83, DerHexer, Derek geary, Devanampriya, Dewritech, Diannaa, Dirrahman, Discospinster, DisillusionedBitterAndKnackered, Dmeranda, Doctordestiny, Dolphin51, Don4of4, Doolotbek e, Drf5n, Drunkenmonkey, Dusk shadow187, Dv82matt, E Wing, EJF, ESkog, Eaglecap Backpack, EdC, Edgar181, Edison, Eequor, Efrainvasquez5th, ElinorD, Enviroboy, Epbr123, Erik Neves, Error -128, Est.r, Ethulin, EugeneZelenko, Eurosong, Everyking, FCYTravis, Fabartus, Farosdaughter, FelisLeo, Fetchcomms, Fieldday-sunday, Flewis, Fluffernutter, FocalPoint, FrankFlanagan, Freeksquad, Full Shunyata, Fæ, Gaff, Gail, Galorr, GhostPirate, Giftlite, Ginsengbomb, GipeDZ, Gisling, Glane23, Glass Sword, Glenn, Gogo Dodo, Grahamperrin, Grimey109, Gun Powder Ma, Gurch, Gwernol, Gwinva, Hadal, Haiduc, Hairy Dude, HamburgerRadio, Hammer1980, Hanchi, Harburg, Harry Bull, Hashim321, Hayabusa future, Hdrhedt, Heqs, Heracles31, Heron, Hgrobe, HiLo48, Hilittledork, Hmrox, Hogyn Lleol, Hookey-rox, Hooperbloob, Horselover Frost, Hottentot, Husond, Hydrargyrum, IRelayer, Icairns, Imjustmatthew, Indy muaddib, Inisheer, Insanity Incarnate, Intranetusa, IntroNats101, Inwind, Iridescent, Irishguy, IronGargoyle, Itai, J.delanoy, J8079s, JForget, Jacobharrisonrichter, Jagged 85, James Lum K.H., JamesR, Jaypee1, JaypeeUS, Jcpenny115632, Jeff G., Jeff3000, Jeltz, Jiddisch, Jim.henderson, Jinian, Jiy, Jlittlenz, JoJaysius, Joconnor, Joel7687, John, John K, John of Reading, JohnOwens, Johnh123, JohnnyJohnny, Jojhutton, Jorgenev, Jose77, Jrvz, Jsmn4vu, Juliancolton, Jusdafax, K Eliza Coyne, KAM, KGasso, Kamal Chid, Kamek98, Kanags, Katalaveno, Katimawan2005, Katpatuka, Keahapana, Keilana, Keycard, KharBevNor, Kingpin13, Kingturtle, Kjl123, Klimot, Knutux, Koavf, Kobalt64, Krish94, Kslotte, Kubigula, Kuru, Kwiki, LOTRrules, La goutte de pluie, Lawrencekhoo, Lazulilasher, Leanwitit, LeaveSleaves, LedgendGamer, Lexi Marie, Lightmouse, LilHelpa, LinDrug, Ling.Nut, Linkspamremover, LjL, Local hero, Logan, Lord Pistachio, LordHarris, Loren Rosen, LtPowers, Luk, Luna Santin, Madden88, Maddie!, Madman2001, Magi42, Maijinsan, Malinaccier, Man vyi, Mani1, Marek69, Mark.murphy, Mark91, Marquez, Martinlc, Master of the Orichalcos, Max Naylor, Maxamegalon2000, Mboverload, Mburgard, McSly, Mekong mainstream, Merlion444, Meters, Methcub, Mgzs180, Michael Daly, Michael Hardy, Michaelkourlas, Mikerom12, Mild Bill Hiccup, Mini-Geek, Ministevieg, Mirwin, Mjsabby, Mlessard, Mlybanon, Mmonne, MoRsE, Modulatum, Monkey Bounce, Morgis, Mspraveen, Mumia-w-18, Musiphil, Mwanner, My76Strat, Mygerardromance, N5iln, NHRHS2010, Naddy, Natalie Erin, Natobxl, NawlinWiki, NeOpets22, NellieBly, NerdyScienceDude, Neyagawa, Nfe, Nigelj, Nikai, Ninthabout, Nk, Nn123645, No Guru, Nommonomanac, NoodleWhacks, Noommos, Nsaa, Nyenyec, O Graeme Burns, Oaktree b, Oceanh, Ohconfucius, Ojigiri, Omg all the good usernames are taken, Onco p53, Oo7565, Orfen, OverlordQ, Oxonhutch, Oxymoron83, Oybubblejohn, PGWG, PMLawrence, Pablomartinez, PaterMcFly, Paul Koning, Paul Mackay, Pax:Vobiscum, Pepper, PericlesofAthens, Petehawk, Peter Ellis, Peter M. Brown, Phanton, Pheifdog, Phgao, Philip Trueman, PhnomPencil, Physis, Piano non troppo, Piepot9, PierreAbbat, PigFlu Oink, Pinethicket, Pit, Polylerus, Pomona17, Populus, Prodego, ProhibitOnions, Python eggs, Qfl247, Quepasahombre, Quinsareth, Qxz, R3venans, RainbowOfLight, Rama, Raryel, RaseaC, Ravedave, Raven in Orbit, Raymondwinn, Reaper Eternal, Reddi, Reinoutr, RevolverOcelotX, RexNL, Rich Farmbrough, Ringomassa, Rjwilmsi, Rkov, Rlhatton, Rob cowie, RockMagnetist, Rogper, Rossami, Rygel, SJP, SajmonDK, Salsa Shark, Salvio giuliano, Santryl, Scientific29, Scriberius, Seb az86556, Secretlondon, Sgt. R.K. Blue, Shantavira, Shavindu Piyadasa, Shreevatsa, Siberex, Sietse Snel, SilkTork, Simetrical, Simon12, Simondrake, Sivullinen, Skizzik, Slashme, Slow Riot, Smalljim, Smelialichu, Snowolfd4, SoHome, Someguy1221, SoniaZ, Sonjaaa, SpK, Spellmaster, Spiko-carpediem, Springnuts, SpuriousQ, Srushe, Ssilvers, Stabilo boss, Stalefries, Stbalbach, Stephenb, Steven Zhang, Suffusion of Yellow, Sureiyah1, Suruena, Sus scrofa, Svm2, Swimtrunkcity, Symane, Syncategoremata, THEN WHO WAS PHONE?, Tetraedycal, The Cunctator, The Flying Spaghetti Monster, The Thing That Should Not Be, The undertow, TheLeopard, TheOtherJesse, Thelettuceman, Thewho98, Thingg, Thrixx, Thumperward, Tide rolls, Tim1357, Tom Pippens, Tom harrison, Tombomp, Tommy2010, Topory, Tpb1093, TransUtopian, Trausten2, Treisijs, Trevor MacInnis, Trjumpet, Trusilver, Tubbs334, Typhoonchaser, Tyranny Sue, Tznkai, UTSRelativity, Udiehte7482, Uditshankar, Ugen64, Uncle Dick, Underorbit, VASANTH S.N., Valfontis, Vary, Velella, Versus22, Vgy7ujm, Wavelength, Wdfarmer, Wetman, WikHead, WikiLaurent, Will O'Neil, William Avery, Williamlear, Wknight94, WongFeiHung, Woohookitty, WookieInHeat, Yamummasnotbad, Yekrats, Zereshk, Zidonuke, Zirguezi, Zootsuits, Zro, Zsinj, Zundark, Zzuuzz, Île flottante, 1381 anonymous edits

Low-frequency_radio_range *Source*: http://en.wikipedia.org/w/index.php?title=Low-frequency_radio_range *Contributors*: Brian in denver, Chris the speller, Crum375, Espoo, Falcon8765, Htonl, Koavf, Malleus Fatuorum, Niceguyedc, Rd232, Sfan00 IMG, 4 anonymous edits

Navigational_aid *Source*: http://en.wikipedia.org/w/index.php?title=Navigational_aid *Contributors*: 7&6=thirteen, Alaniaris, Alansohn, Beeblebrox, Bullestock, Carl.bunderson, Cdc, Cuprum17, DGPS, Dual Freq, Fayenatic london, Haus, John Broughton, Krellis, LPFixit, Mouldus, Muhandes, Neutrality, Peter Ellis, Safemariner, Ser Amantio di Nicolao, Skapur, Tinton5, Uruiamme, West.andrew.g, Wikiedit567, 12 anonymous edits

1946_Australian_National_Airways_DC-3_crash *Source*: http://en.wikipedia.org/w/index.php?title=1946_Australian_National_Airways_DC-3_crash *Contributors*: DexDor, Dolphin51, Grandiose, Karlap, Mdnavman, Meltdown627, MilborneOne, Petebutt

Image Sources, Licenses and Contributors

Image:Douglas DC-3, SE-CFP.jpg *Source*: http://en.wikipedia.org/w/index.php?title=File:Douglas_DC-3,_SE-CFP.jpg *License*: unknown *Contributors*: Towpilot

Image:dc3.takeoff.thales.arp.jpg *Source*: http://en.wikipedia.org/w/index.php?title=File:Dc3.takeoff.thales.arp.jpg *License*: unknown *Contributors*: Ardfern, Arpingstone, El Grafo, High Contrast, KTo288, Timak, Tony Wills, 2 anonymous edits

Image:Ntps-c47-N834TP-071112-01cr-16.jpg *Source*: http://en.wikipedia.org/w/index.php?title=File:Ntps-c47-N834TP-071112-01cr-16.jpg *License*: unknown *Contributors*: User:Akradecki

Image:Douglas DC-3 Aigle Azur Palas Jet 1953.jpg *Source*: http://en.wikipedia.org/w/index.php?title=File:Douglas_DC-3_Aigle_Azur_Palas_Jet_1953.jpg *License*: unknown *Contributors*: User:RuthAS

Image:DC-3 on Floats N130Q.JPG *Source*: http://en.wikipedia.org/w/index.php?title=File:DC-3_on_Floats_N130Q.JPG *License*: unknown *Contributors*: TimothyMN (talk). Original uploader was TimothyMN at en.wikipedia

File:1 free nz photo dc-3.jpg *Source*: http://en.wikipedia.org/w/index.php?title=File:1_free_nz_photo_dc-3.jpg *License*: unknown *Contributors*: User:QFSE Media

Image:DC-3 in SoAfrica.jpg *Source*: http://en.wikipedia.org/w/index.php?title=File:DC-3_in_SoAfrica.jpg *License*: unknown *Contributors*: Original uploader was Doontass at en.wikipedia

Image:DC3 of Indigo Airlines at Pemba Airport Aug 2009.jpg *Source*: http://en.wikipedia.org/w/index.php?title=File:DC3_of_Indigo_Airlines_at_Pemba_Airport_Aug_2009.jpg *License*: unknown *Contributors*: User:Trekkwik

File:Fujairah Airlines Douglas DC-3 Wheatley.jpg *Source*: http://en.wikipedia.org/w/index.php?title=File:Fujairah_Airlines_Douglas_DC-3_Wheatley.jpg *License*: unknown *Contributors*: John M. Wheatley

File:Nakajima L2D2 at Zamboanga 1945.jpeg *Source*: http://en.wikipedia.org/w/index.php?title=File:Nakajima_L2D2_at_Zamboanga_1945.jpeg *License*: unknown *Contributors*: U.S. Navy

File:Maritime Patrol and Rescue Conroy Tri-Turbo Three Fitzgerald.jpg *Source*: http://en.wikipedia.org/w/index.php?title=File:Maritime_Patrol_and_Rescue_Conroy_Tri-Turbo_Three_Fitzgerald.jpg *License*: unknown *Contributors*: Steve Fitzgerald

Image:DC3 Silh.jpg *Source*: http://en.wikipedia.org/w/index.php?title=File:DC3_Silh.jpg *License*: unknown *Contributors*: Crown copyright

Image:N34---Douglas-DC3-Cockpit.jpg *Source*: http://en.wikipedia.org/w/index.php?title=File:N34---Douglas-DC3-Cockpit.jpg *License*: unknown *Contributors*: Intersofia, KTo288, PMG, Rcbutcher, Stahlkocher, , 3 anonymous edits

File:B7279.jpg *Source*: http://en.wikipedia.org/w/index.php?title=File:B7279.jpg *License*: unknown *Contributors*: Wongm

File:Nla-pic-an24056977-v.jpeg *Source*: http://en.wikipedia.org/w/index.php?title=File:Nla-pic-an24056977-v.jpeg *License*: unknown *Contributors*: Moondyne, Wongm

File:012059d.jpg *Source*: http://en.wikipedia.org/w/index.php?title=File:012059d.jpg *License*: unknown *Contributors*: Original uploader was Wongm at en.wikipedia. Later version(s) were uploaded by Moondyne at en.wikipedia.

File:Canberra COA.gif *Source*: http://en.wikipedia.org/w/index.php?title=File:Canberra_COA.gif *License*: unknown *Contributors*: BeatrixBelibaste, Cflm001, Secar one

File:Flag of the Australian Capital Territory.svg *Source*: http://en.wikipedia.org/w/index.php?title=File:Flag_of_the_Australian_Capital_Territory.svg *License*: unknown *Contributors*: user:Dbenbenn

File:Flag of Australia.svg *Source*: http://en.wikipedia.org/w/index.php?title=File:Flag_of_Australia.svg *License*: unknown *Contributors*: Anomie, Mifter

File:Law Courts of the Australian Capital Territory.jpg *Source*: http://en.wikipedia.org/w/index.php?title=File:Law_Courts_of_the_Australian_Capital_Territory.jpg *License*: unknown *Contributors*: User:Bidgee

File:Kompas Sofia.JPG *Source*: http://en.wikipedia.org/w/index.php?title=File:Kompas_Sofia.JPG *License*: unknown *Contributors*: Bios, Homonihilis, Pymouss

File:Smartphone Compass.jpg *Source*: http://en.wikipedia.org/w/index.php?title=File:Smartphone_Compass.jpg *License*: unknown *Contributors*: User:Zirguezi

File:Model Si Nan of Han Dynasty.jpg *Source*: http://en.wikipedia.org/w/index.php?title=File:Model_Si_Nan_of_Han_Dynasty.jpg *License*: unknown *Contributors*: HéctorTabaré, Pymouss, Typo, Yug, 2 anonymous edits

File:Ming-marine-compass.jpg *Source*: http://en.wikipedia.org/w/index.php?title=File:Ming-marine-compass.jpg *License*: unknown *Contributors*: Gisling, Man vyi, Stunteltje, 2 anonymous edits

File:Epistola-de-magnete.jpg *Source*: http://en.wikipedia.org/w/index.php?title=File:Epistola-de-magnete.jpg *License*: unknown *Contributors*: Louis le Grand, Pymouss

File:Compass thumbnail.jpg *Source*: http://en.wikipedia.org/w/index.php?title=File:Compass_thumbnail.jpg *License*: unknown *Contributors*: It Is Me Here, Liftarn, Mattes, Psychonaut, Roomba, Schimmelreiter, Wst, 2 anonymous edits

File:Kardanischer-Kompass.jpg *Source*: http://en.wikipedia.org/w/index.php?title=File:Kardanischer-Kompass.jpg *License*: unknown *Contributors*: Louis le Grand, Pymouss, Santosga

File:Bearing compass.jpg *Source*: http://en.wikipedia.org/w/index.php?title=File:Bearing_compass.jpg *License*: unknown *Contributors*: User:Inisheer

File:Flush mount compass.jpg *Source*: http://en.wikipedia.org/w/index.php?title=File:Flush_mount_compass.jpg *License*: unknown *Contributors*: User:SoHome

File:Aero Magnetic Compass.jpg *Source*: http://en.wikipedia.org/w/index.php?title=File:Aero_Magnetic_Compass.jpg *License*: unknown *Contributors*: Chopper

File:walkers compass arp.jpg *Source*: http://en.wikipedia.org/w/index.php?title=File:Walkers_compass_arp.jpg *License*: unknown *Contributors*: Arpingstone, Pieter Kuiper, Pymouss

File:liquid filled compass.jpg *Source*: http://en.wikipedia.org/w/index.php?title=File:Liquid_filled_compass.jpg *License*: unknown *Contributors*: Nikai, Pymouss

File:Cammenga-lensatic-compass-model-27.jpg *Source*: http://en.wikipedia.org/w/index.php?title=File:Cammenga-lensatic-compass-model-27.jpg *License*: unknown *Contributors*: User:Siberex

File:Compasses orienteering.jpg *Source*: http://en.wikipedia.org/w/index.php?title=File:Compasses_orienteering.jpg *License*: unknown *Contributors*: User:Leinad

File:Brunton.JPG *Source*: http://en.wikipedia.org/w/index.php?title=File:Brunton.JPG *License*: unknown *Contributors*: User:Citypeek

File:Boussole fantassin russe.jpg *Source*: http://en.wikipedia.org/w/index.php?title=File:Boussole_fantassin_russe.jpg *License*: unknown *Contributors*: Jean-Patrick Donzey

File:MuseeMarine-compas-p1000468.jpg *Source*: http://en.wikipedia.org/w/index.php?title=File:MuseeMarine-compas-p1000468.jpg *License*: unknown *Contributors*: User:Rama

File:CompassUseMapMarked.jpg *Source*: http://en.wikipedia.org/w/index.php?title=File:CompassUseMapMarked.jpg *License*: unknown *Contributors*: Audrius Meskauskas

File:CompassUseTargetMarked.jpg *Source*: http://en.wikipedia.org/w/index.php?title=File:CompassUseTargetMarked.jpg *License*: unknown *Contributors*: Audrius Meskauskas

File:Measuring azimuth with a compass.jpg *Source*: http://en.wikipedia.org/w/index.php?title=File:Measuring_azimuth_with_a_compass.jpg *License*: unknown *Contributors*: US WAR DEPARTMENT, UNITED STATES GOVERNMENT PRINTING OFFICE WASHINGTON : 1941

File:A-N-signals.png *Source*: http://en.wikipedia.org/w/index.php?title=File:A-N-signals.png *License*: unknown *Contributors*: Crum375 (talk). Original uploader was Crum375 at en.wikipedia

File:Lt. General James Doolittle, head and shoulders.jpg *Source*: http://en.wikipedia.org/w/index.php?title=File:Lt._General_James_Doolittle,_head_and_shoulders.jpg *License*: unknown *Contributors*: BrokenSphere, Nobunaga24, Rmhermen, Tom

File:Doolittle-inst-panel.jpg *Source*: http://en.wikipedia.org/w/index.php?title=File:Doolittle-inst-panel.jpg *License*: unknown *Contributors*: United States Air Force

File:LFR-reed.jpg *Source*: http://en.wikipedia.org/w/index.php?title=File:LFR-reed.jpg *License*: unknown *Contributors*: United States Air Force

File:LFR-loop.jpg *Source*: http://en.wikipedia.org/w/index.php?title=File:LFR-loop.jpg *License*: unknown *Contributors*: United States Air Force

File:LFR-photo.jpg *Source*: http://en.wikipedia.org/w/index.php?title=File:LFR-photo.jpg *License*: unknown *Contributors*: U.S. DOT FAA

File:4-course.png *Source*: http://en.wikipedia.org/w/index.php?title=File:4-course.png *License*: unknown *Contributors*: U.S. Coast and Geodetic Survey (NOAA)

File:JOT-LFR.png *Source*: http://en.wikipedia.org/w/index.php?title=File:JOT-LFR.png *License*: unknown *Contributors*: U.S. Dept. of Commerce, Coast and Geodetic Survey. Original uploader was Crum375 at en.wikipedia

File:LFR-hold.jpg *Source*: http://en.wikipedia.org/w/index.php?title=File:LFR-hold.jpg *License*: unknown *Contributors*: United States Air Force

File:Nkr1.jpg *Source*: http://en.wikipedia.org/w/index.php?title=File:Nkr1.jpg *License*: unknown *Contributors*: Original uploader was Mintaka10 at en.wikipedia Later version(s) were uploaded by Hydrargyrum, Zonk43 at en.wikipedia.

File:Table Rock VOR.jpg *Source*: http://en.wikipedia.org/w/index.php?title=File:Table_Rock_VOR.jpg *License*: unknown *Contributors*: ZabMilenko

Image:NobskaLight.jpg *Source*: http://en.wikipedia.org/w/index.php?title=File:NobskaLight.jpg *License*: unknown *Contributors*: Jameslwoodward, Jonesey, Julo, Mtsmallwood

Image:ATON - Road Signs of the Waterway.gif *Source*: http://en.wikipedia.org/w/index.php?title=File:ATON_-_Road_Signs_of_the_Waterway.gif *License*: unknown *Contributors*: United States Coast Guard

GNU Free Documentation License Version 1.2, November 2002 Copyright (C) 2000,2001,2002 Free Software Foundation, Inc. 59 Temple Place, Suite 330, Boston, MA 02111-1307 USA Everyone is permitted to copy and distribute verbatim copies of this license document, but changing it is not allowed.

0. PREAMBLE

The purpose of this License is to make a manual, textbook, or other functional and useful document "free" in the sense of freedom: to assure everyone the effective freedom to copy and redistribute it, with or without modifying it, either commercially or noncommercially. Secondarily, this License preserves for the author and publisher a way to get credit for their work, while not being considered responsible for modifications made by others. This License is a kind of "copyleft", which means that derivative works of the document must themselves be free in the same sense. It complements the GNU General Public License, which is a copyleft license designed for free software. We have designed this License in order to use it for manuals for free software, because free software needs free documentation: a free program should come with manuals providing the same freedoms that the software does. But this License is not limited to software manuals; it can be used for any textual work, regardless of subject matter or whether it is published as a printed book. We recommend this License principally for works whose purpose is instruction or reference.

1. APPLICABILITY AND DEFINITIONS

This License applies to any manual or other work, in any medium, that contains a notice placed by the copyright holder saying it can be distributed under the terms of this License. Such a notice grants a world-wide, royalty-free license, unlimited in duration, to use that work under the conditions stated herein. The "Document", below, refers to any such manual or work. Any member of the public is a licensee, and is addressed as "you". You accept the license if you copy, modify or distribute the work in a way requiring permission under copyright law. A "Modified Version" of the Document means any work containing the Document or a portion of it, either copied verbatim, or with modifications and/or translated into another language. A "Secondary Section" is a named appendix or a front-matter section of the Document that deals exclusively with the relationship of the publishers or authors of the Document to the Document's overall subject (or to related matters) and contains nothing that could fall directly within that overall subject. (Thus, if the Document is in part a textbook of mathematics, a Secondary Section may not explain any mathematics.) The relationship could be a matter of historical connection with the subject or with related matters, or of legal, commercial, philosophical, ethical or political position regarding them. The "Invariant Sections" are certain Secondary Sections whose titles are designated, as being those of Invariant Sections, in the notice that says that the Document is released under this License. If a section does not fit the above definition of Secondary then it is not allowed to be designated as Invariant. The Document may contain zero Invariant Sections. If the Document does not identify any Invariant Sections then there are none. The "Cover Texts" are certain short passages of text that are listed, as Front-Cover Texts or Back-Cover Texts, in the notice that says that the Document is released under this License. A Front-Cover Text may be at most 5 words, and a Back-Cover Text may be at most 25 words. A "Transparent" copy of the Document means a machine-readable copy, represented in a format whose specification is available to the general public, that is suitable for revising the document straightforwardly with generic text editors or (for images composed of pixels) generic paint programs or (for drawings) some widely available drawing editor, and that is suitable for input to text formatters or for automatic translation to a variety of formats suitable for input to text formatters. A copy made in an otherwise Transparent file format whose markup, or absence of markup, has been arranged to thwart or discourage subsequent modification by readers is not Transparent. An image format is not Transparent if used for any substantial amount of text. A copy that is not "Transparent" is called "Opaque". Examples of suitable formats for Transparent copies include plain ASCII without markup, Texinfo input format, LaTeX input format, SGML or XML using a publicly available DTD, and standard-conforming simple HTML, PostScript or PDF designed for human modification. Examples of transparent image formats include PNG, XCF and JPG. Opaque formats include proprietary formats that can be read and edited only by proprietary word processors, SGML or XML for which the DTD and/or processing tools are not generally available, and the machine-generated HTML, PostScript or PDF produced by some word processors for output purposes only. The "Title Page" means, for a printed book, the title page itself, plus such following pages as are needed to hold, legibly, the material this License requires to appear in the title page. For works in formats which do not have any title page as such, "Title Page" means the text near the most prominent appearance of the work's title, preceding the beginning of the body of the text. A section "Entitled XYZ" means a named subunit of the Document whose title either is precisely XYZ or contains XYZ in parentheses following text that translates XYZ in another language. (Here XYZ stands for a specific section name mentioned below, such as "Acknowledgements", "Dedications", "Endorsements", or "History".) To "Preserve the Title" of such a section when you modify the Document means that it remains a section "Entitled XYZ" according to this definition. The Document may include Warranty Disclaimers next to the notice which states that this License applies to the Document. These Warranty Disclaimers are considered to be included by reference in this License, but only as regards disclaiming warranties: any other implication that these Warranty Disclaimers may have is void and has no effect on the meaning of this License.

2. VERBATIM COPYING

You may copy and distribute the Document in any medium, either commercially or noncommercially, provided that this License, the copyright notices, and the license notice saying this License applies to the Document are reproduced in all copies, and that you add no other conditions whatsoever to those of this License. You may not use technical measures to obstruct or control the reading or further copying of the copies you make or distribute. However, you may accept compensation in exchange for copies. If you distribute a large enough number of copies you must also follow the conditions in section 3. You may also lend copies, under the same conditions stated above, and you may publicly display copies.

3. COPYING IN QUANTITY

If you publish printed copies (or copies in media that commonly have printed covers) of the Document, numbering more than 100, and the Document's license notice requires Cover Texts, you must enclose the copies in covers that carry, clearly and legibly, all these Cover Texts: Front-Cover Texts on the front cover, and Back-Cover Texts on the back cover. Both covers must also clearly and legibly identify you as the publisher of these copies. The front cover must present the full title with all words of the title equally prominent and visible. You may add other material on the covers in addition. Copying with changes limited to the covers, as long as they preserve the title of the Document and satisfy these conditions, can be treated as verbatim copying in other respects. If the required texts for either cover are too voluminous to fit legibly, you should put the first ones listed (as many as fit reasonably) on the actual cover, and continue the rest onto adjacent pages. If you publish or distribute Opaque copies of the Document numbering more than 100, you must either include a machine-readable Transparent copy along with each Opaque copy, or state in or with each Opaque copy a computer-network location from which the general network-using public has access to download using public-standard network protocols a complete Transparent copy of the Document, free of added material. If you use the latter option, you must take reasonably prudent steps, when you begin distribution of Opaque copies in quantity, to ensure that this Transparent copy will remain thus accessible at the stated location until at least one year after the last time you distribute an Opaque copy (directly or through your agents or retailers) of that edition to the public. It is requested, but not required, that you contact the authors of the Document well before redistributing any large number of copies, to give them a chance to provide you with an updated version of the Document.

4. MODIFICATIONS

You may copy and distribute a Modified Version of the Document under the conditions of sections 2 and 3 above, provided that you release the Modified Version under precisely this License, with the Modified Version filling the role of the Document, thus licensing distribution and modification of the Modified Version to whoever possesses a copy of it. In addition, you must do these things in the Modified Version: A. Use in the Title Page (and on the covers, if any) a title distinct from that of the Document, and from those of previous versions (which should, if there were any, be listed in the History section of the Document). You may use the same title as a previous version if the original publisher of that version gives permission. B. List on the Title Page, as authors, one or more persons or entities responsible for authorship of the modifications in the Modified Version, together with at least five of the principal authors of the Document (all of its principal authors, if it has fewer than five), unless they release you from this requirement. C. State on the Title page the name of the publisher of the Modified Version, as the publisher. D. Preserve all the copyright notices of the Document. E. Add an appropriate copyright notice for your modifications adjacent to the other copyright notices. F. Include, immediately after the copyright notices, a license notice giving the public permission to use the Modified Version under the terms of this License, in the form shown in the Addendum below. G. Preserve in that license notice the full lists of Invariant Sections and required Cover Texts given in the Document's license notice. H. Include an unaltered copy of this License. I. Preserve the section Entitled "History", Preserve its Title, and add to it an item stating at least the title, year, new authors, and publisher of the Modified Version as given on the Title Page. If there is no section Entitled "History" in the Document, create one stating the title, year, authors, and publisher of the Document as given on its Title Page, then add an item describing the Modified Version as stated in the previous sentence. J. Preserve the network location, if any, given in the Document for public access to a Transparent copy of the Document, and likewise the network locations given in the Document for previous versions it was based on. These may be placed in the "History" section. You may omit a network location for a work that was published at least four years before the Document itself, or if the original publisher of the version it refers to gives permission. K. For any section Entitled "Acknowledgements" or "Dedications", Preserve the Title of the section, and preserve in the section all the substance and tone of each of the contributor acknowledgements and/or dedications given therein. L. Preserve all the Invariant Sections of the Document, unaltered in their text and in their titles. Section numbers or the equivalent are not considered part of the section titles. M. Delete any section Entitled "Endorsements". Such a section may not be included in the Modified Version. N. Do not retitle any existing section to be Entitled "Endorsements" or to conflict in title with any Invariant Section. O. Preserve any Warranty Disclaimers. If the Modified Version includes new front-matter sections or appendices that qualify as Secondary Sections and contain no material copied from the Document, you may at your option designate some or all of these sections as invariant. To do this, add their titles to the list of Invariant Sections in the Modified Version's license notice. These titles must be distinct from any other section titles. You may add a section Entitled "Endorsements", provided it contains nothing but endorsements of your Modified Version by various parties--for example, statements of peer review or that the text has been approved by an organization as the authoritative definition of a standard. You may add a passage of up to five words as a Front-Cover Text, and a passage of up to 25 words as a Back-Cover Text, to the end of the list of Cover Texts in the Modified Version. Only one passage of Front-Cover Text and one of Back-Cover Text may be added by (or through arrangements made by) any one entity. If the Document already includes a cover text for the same cover, previously added by you or by arrangement made by the same entity you are acting on behalf of, you may not add another; but you may replace the old one, on explicit permission from the previous publisher that added the old one. The author(s) and publisher(s) of the Document do not by this License give permission to use their names for publicity for or to assert or imply endorsement of any Modified Version.

5. COMBINING DOCUMENTS

You may combine the Document with other documents released under this License, under the terms defined in section 4 above for modified versions, provided that you include in the combination all of the Invariant Sections of all of the original documents, unmodified, and list them all as Invariant Sections of your combined work in its license notice, and that you preserve all their Warranty Disclaimers. The combined work need only contain one copy of this License, and multiple identical Invariant Sections may be replaced with a single copy. If there are multiple Invariant Sections with the same name but different contents, make the title of each such section unique by adding at the end of it, in parentheses, the name of the original author or publisher of that section if known, or else a unique number. Make the same adjustment to the section titles in the list of Invariant Sections in the license notice of the combined work. In the combination, you must combine any sections Entitled "History" in the various original documents, forming one section Entitled "History"; likewise combine any sections Entitled "Acknowledgements", and any sections Entitled "Dedications". You must delete all sections Entitled "Endorsements".

6. COLLECTIONS OF DOCUMENTS

You may make a collection consisting of the Document and other documents released under this License, and replace the individual copies of this License in the various documents with a single copy that is included in the collection, provided that you follow the rules of this License for verbatim copying of each of the documents in all other respects. You may extract a single document from such a collection, and distribute it individually under this License, provided you insert a copy of this License into the extracted document, and follow this License in all other respects regarding verbatim copying of that document.

7. AGGREGATION WITH INDEPENDENT WORKS

A compilation of the Document or its derivatives with other separate and independent documents or works, in or on a volume of a storage or distribution medium, is called an "aggregate" if the copyright resulting from the compilation is not used to limit the legal rights of the compilation's users beyond what the individual works permit. When the Document is included in an aggregate, this License does not apply to the other works in the aggregate which are not themselves derivative works of the Document. If the Cover Text requirement of section 3 is applicable to these copies of the Document, then if the Document is less than one half of the entire aggregate, the Document's Cover Texts may be placed on covers that bracket the Document within the aggregate, or the electronic equivalent of covers if the Document is in electronic form. Otherwise they must appear on printed covers that bracket the whole aggregate.

8. TRANSLATION

Translation is considered a kind of modification, so you may distribute translations of the Document under the terms of section 4. Replacing Invariant Sections with translations requires special permission from their copyright holders, but you may include translations of some or all Invariant Sections in addition to the original versions of these Invariant Sections. You may include a translation of this License, and all the license notices in the Document, and any Warranty Disclaimers, provided that you also include the original English version of this License and the original versions of those notices and disclaimers. In case of a disagreement between the translation and the original version of this License or a notice or disclaimer, the original version will prevail. If a section in the Document is Entitled "Acknowledgements", "Dedications", or "History", the requirement (section 4) to Preserve its Title (section 1) will typically require changing the actual title.

9. TERMINATION

You may not copy, modify, sublicense, or distribute the Document except as expressly provided for under this License. Any other attempt to copy, modify, sublicense or distribute the Document is void, and will automatically terminate your rights under this License. However, parties who have received copies, or rights, from you under this License will not have their licenses terminated so long as such parties remain in full compliance.

10. FUTURE REVISIONS OF THIS LICENSE

The Free Software Foundation may publish new, revised versions of the GNU Free Documentation License from time to time. Such new versions will be similar in spirit to the present version, but may differ in detail to address new problems or concerns. See http://www.gnu.org/copyleft/. Each version of the License is given a distinguishing version number. If the Document specifies that a particular numbered version of this License "or any later version" applies to it, you have the option of following the terms and conditions either of that specified version or of any later version that has been published (not as a draft) by the Free Software Foundation. If the Document does not specify a version number of this License, you may choose any version ever published (not as a draft) by the Free Software Foundation. ADDENDUM: How to use this License for your documents To use this License in a document you have written, include a copy of the License in the document and put the following copyright and license notices just after the title page: Copyright (c) YEAR YOUR NAME. Permission is granted to copy, distribute and/or modify this document under the terms of the GNU Free Documentation License, Version 1.2 or any later version published by the Free Software Foundation; with no Invariant Sections, no Front-Cover Texts, and no Back-Cover Texts. A copy of the license is included in the section entitled "GNU Free Documentation License". If you have Invariant Sections, Front-Cover Texts and Back-Cover Texts, replace the "with...Texts." line with this: with the Invariant Sections being LIST THEIR TITLES, with the Front-Cover Texts being LIST, and with the Back-Cover Texts being LIST. If you have Invariant Sections without Cover Texts, or some other combination of the three, merge those two alternatives to suit the situation. If your document contains nontrivial examples of program code, we recommend releasing these examples in parallel under your choice of free software license, such as the GNU General Public License, to permit their use in free software.

Printed by Books on Demand GmbH, Norderstedt / Germany